T0135808

Bibliografische Information der Deutschen Nationalbibliothek

Die Deutsche Nationalbibliothek verzeichnet diese Publikation in der
Deutschen Nationalbibliografie; detaillierte bibliografische Daten sind
im Internet über http://dnb.d-nb.de abrufbar.

ISBN 978-3-8325-3297-0

Logos Verlag Berlin GmbH
Comeniushof, Gubener Str. 47,
10243 Berlin
Tel.: +49 (0)30 42 85 10 90
Fax: +49 (0)30 42 85 10 92
INTERNET: http://www.logos-verlag.de

Conceptual Motorics – Generation and Evaluation of Communicative Robot Gesture

by

Maha Salem

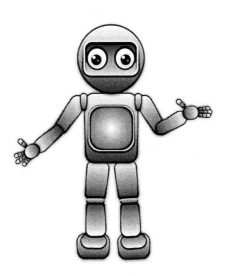

Conceptual Motorics – Generation and Evaluation of Communicative
Robot Gesture

Dissertation submitted to Bielefeld University in partial fulfillment of the requirements for the degree of "Doctor of Engineering" (Dr.-Ing.)

Maha Salem
Research Institute for Cognition and Robotics
Faculty of Technology
Bielefeld University
P.O. Box 10 01 31
D-33501 Bielefeld
Germany
Email: msalem@techfak.uni-bielefeld.de

DEAN OF THE FACULTY:
Prof. Dr. Jens Stoye

THESIS REVIEWERS:
Prof. Dr.-Ing. Stefan Kopp
Dr.-Ing. Frank Joublin
Prof. Dr. Vanessa Evers

EXAMINATION COMMITTEE:
Prof. Dr.-Ing. Ulrich Rückert
Prof. Dr.-Ing. Stefan Kopp
Dr.-Ing. Frank Joublin
Prof. Dr. Vanessa Evers
Dr. Manja Lohse

Date of Thesis Submission: 18 April 2012
Date of Thesis Defense: 4 July 2012

Abstract

How do humans perceive communicative gesture behavior in robots? Although gesture is a crucial feature of social interaction, this research question is still largely unexplored in the field of social robotics. The present work thus sets out to investigate how robot gesture can be used to design and realize more natural and human-like communication capabilities for social robots. The adopted approach is twofold. Firstly, the technical challenges encountered when implementing a speech-gesture generation model on a robotic platform are addressed. The realized framework enables a humanoid robot to produce synthetic speech and co-verbal hand and arm gestures. In contrast to many existing generation models, these gestures are not limited to a predefined repertoire of motor actions but are flexibly generated at run-time. Fine synchronization of the two modalities is achieved by means of a sophisticated multimodal scheduler specifically implemented for humanoid robot gesture and speech. Secondly, the achieved expressiveness and flexibility in robot gesture is exploited in controlled experiments. To gain a deeper understanding of how communicative robot gesture might impact and shape human perception and evaluation of human-robot interaction, two experimental studies were conducted. The findings reveal that participants evaluate the robot more positively when non-verbal behaviors such as hand and arm gestures are displayed along with speech. Surprisingly, this effect was particularly pronounced when the robot's gesturing behavior was partly incongruent with speech. These findings contribute new insights into human perception and understanding of non-verbal behaviors in artificial embodied agents. Ultimately, they support the presented approach of endowing social robots with communicative gesture.

Keywords: Multimodal Interaction and Conversational Skills, Robot Gesture, Non-verbal Cues and Expressiveness, Social Human-Robot Interaction, Robot Companions and Social Robots.

Acknowledgements

One month before I was due to start with my PhD, my friend Alex who had just finished his tried his best to persuade me *not* to do it. Of course I didn't follow his advice – or I wouldn't be writing these lines now – but there were many moments during the past few years at which I almost wished I had. The work described in this thesis would not have been possible without the many people who supported me and believed in me throughout what has been the toughest period of my life.

First, I would like to thank my principal supervisor Stefan Kopp for his guidance, advice, and encouragement, which helped me become the "grown-up" researcher that I am now. Besides sparking my interest in gesture research, he taught me not to pursue the easy route but to think critically and accept new challenges as they arise. Second, I am very grateful to Ipke Wachsmuth for his support and friendly advice as my co-supervisor throughout the years of my PhD studies. His occasional, rather unconventional suggestion to take a break from work and go out with friends or to the cinema instead was particularly appreciated! Another big thank you goes to Katharina Rohlfing who has been an exceptional advisor and mentor during my PhD. I came to deeply value both her professional and emotional support that more than once helped me see light in moments of apparent darkness. I am also very grateful to Frank Joublin for being a fantastic external advisor and for the great support during my research stays at the Honda Research Institute Europe. I appreciate the time he put into my thesis review and into the many publication proposals which he had to process on my behalf - I promise there won't be too many more in the near future! I would further like to thank Vanessa Evers for kindly agreeing to be an external examiner in my thesis committee and for the time and effort associated with this commitment.

I'd like to thank my amazing colleagues for the frequent reminder that, despite my work on human-robot interaction, I actually enjoy human-human interaction more than I jokingly admit sometimes. Many thanks go to the people from the CoR-Lab – especially the "N5-Gang" including Christian Emmerich, Raphael Golombek, Christian Lang, Andre Lemme, Heiko Lex, Ananda Freire, Klaus Neumann, Arne Nordmann, Felix Reinhart, Matthias Rolf, Stefan Rüther, Andrea Soltoggio (who also kindly provided me with this fancy LaTeX template for my thesis), and Johannes Wienke – for the excellent work environment, the cheerful coffee breaks, the football tournaments, and all the other countless good memories I will keep from our time working together. Special thanks go to Raphael for being a fantastic officemate who provided me with my daily dose of sarcasm and even tolerated my possibly slightly 'quirky' office decor. Many thanks also go to Lars Schillingmann for being the shell-script fanatic whose helpfulness often resulted in endless discussions in which I would question his methods but which – I hate to admit – typically led to good solutions nonetheless! I would also like to thank Carola Haumann, Anke Kloock, Ruth Moradbakhti, and Jochen Steil for the great support at the administrative level and beyond. I am further grateful to Sebastian Gieselmann, Michael Götting, Stefan Krüger, and Oliver Lieske for their support in the robot lab while I was conducting my experimental studies.

I wish to thank the members of the Sociable Agents Group, including Kirsten Bergmann, Elmar Bienek, Hendrik Buschmeier, Ulf Großekathöfer, Dagmar Philipp, Sebastian Ptock, Amir Sadeghipour, and Ramin Yaghoubzadeh, for the stimulating and friendly work environment. I would further like to thank the members of the Artificial Intelligence Group for always making me feel "at home" in their group too. Special thanks go to Margret Barner for her kindness and unconditional help on numerous occasions, and to Hana Boukricha, Alexa Breuing, and Nhung Nguyen for the friendships and good times we shared beyond PhD work. I'm also grateful to Friederike Eyssel for her help with the data analysis. Many thanks go to the people from the Honda Research Institute Europe, especially to Michael Gienger and Manuel Mühlig, for the kind help and cooperation, and for always making me feel very welcome during my research visits there.

My life as a PhD student would have been much less enjoyable without the wonderful friends who, by occasionally reminding me that there is actually a world outside the lab, probably prevented me from turning into a robot myself. I'd like to thank Alex Campbell for warning me of PhD hell – now I know what he was talking about, and I'll definitely listen to him next time I consider doing a PhD! Thanks to Dennis Wiebusch for the unforgettable "nerd sessions", to Alan Woodley for being an incredibly supportive friend and for not giving up on me while I had disappeared in my "cave" for several years, and to Henning Worm for forcing me to get out of the lab and have a coffee break whenever I was about to lose my mind. Thanks to many more who were there for me in one way or another – Karen Brandt, Yeu-Ying Cheng, Therese Milanovic, Anja & Chris Nagrit, Niko Prüßner, Breanna Studenka, Heidi Williamson, Yasemin Yazar, ... – for their encouragement, support, and friendship throughout and beyond the years of my PhD. I'm extremely grateful to Ben Jones for proofreading a full-length manuscript of my thesis, resulting in many helpful comments and suggestions, and for his remarkable kindness and support as a friend. A huge thank you further goes to Cora Herold for being the most amazing best friend I could dream of.

Finally, I am deeply grateful to my parents, Amal and Sami, and my siblings, Marwa and Ahmed, for their endless support, incredible kindness, and constant belief in me – and for simply being awesome.

iv

Contents

CONTENTS

List of Figures

List of Tables

Chapter 1

Introduction

1.1 Motivation

Robots today are very different than they were about 50 years ago. While industrial robots with functional layouts and production-specific purposes were dominating in the late 1950s, they are no longer designed to function exclusively as manufacturing aids. Remarkable advances made since the introduction of the first robots have led to a great diversity of robotics applications as well as mechanical designs. The wide range of robotics applications today includes museum and reception attendants, toys and entertainment devices, household and service robots, route guides, educational robots, and robots for elderly assistance, therapy, and rehabilitation.

In light of such advances, the roles of robots have become increasingly social, thus bringing about a shift from machines that are designed for traditional human-robot interaction (HRI), such as tele-operated mechanical manipulation, to robots intended for social HRI. In view of this transformation, many authors have demanded improved design for robots to be capable of engaging in meaningful social interactions with humans (e.g., Breazeal, 2003). Along with the attempt to define and name this new category of robots – with Fong et al. (2003) calling them "socially interactive robots" and Breazeal (2002) using the term "sociable robots" – a whole new research area has since emerged.

Social robotics research is dedicated to designing, developing and evaluating robots that can engage in social environments in a way that is appealing to human interaction partners. However, interaction is often difficult because inexperienced users do not understand the robot's internal states, intentions, actions, and expectations. To facilitate successful interaction, social robots should provide communicative functionality that is both natural and intuitive to humans. The appropriate level of such communicative functionality strongly depends on the appearance of the robot and attributions thus made to it. Different design approaches can be chosen depending on the social context and use of the robot. Fong et al. (2003) define four broad categories of social robots based on their appearance and level of embodiment: *anthropomorphic, zoomorphic, caricatured,* and *functionally designed* robots (see **Figure 1.1** for illustrative examples).

While the last three design categories are targeted at establishing a human-creature relationship which does not evoke as high an expectation on the human's side, anthropomorphic design, in contrast, is broadly recommended to support an intuitive and meaningful interaction with humans (Breazeal, 2002; Duffy, 2003). Furthermore, equipping the robot with human-like body features, such as a head, two arms, and two legs, is considered a useful means to elicit the broad spectrum of responses that humans usually direct toward one another. This phenomenon is typically referred to as *anthropomorphism* (Epley et al., 2007), i.e., the attribution of human qualities to non-living objects, and it is increased when "social" movements or behaviors are displayed by the robot (Duffy, 2003). But what types of social movements or behaviors can lead to an increased acceptance of robot companions based on such anthropomorphic inferences?

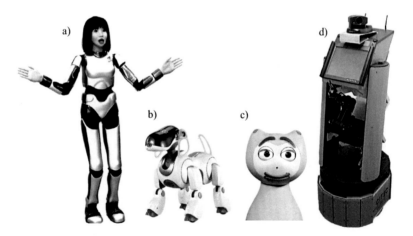

Figure 1.1: Four basic categories and according examples of social robots based on appearance and level of embodiment as defined by Fong et al. (2003): a) *anthropomorphic* (AIST's HRP-4C), b) *zoomorphic* (Sony's Aibo), c) *caricatured* (Philips' iCat), d) *functionally designed* (Bielefeld University's BIRON).

Fong et al. (2003) identify the use of gestures as one crucial aspect when designing robots that are intended to engage in meaningful social interactions with humans. In fact, given the design of humanoid robots, they are typically expected to exhibit human-like communicative behaviors, using their bodies for non-verbal expression just as humans do. Representing an integral component of human communicative behavior (Kendon, 1986; McNeill, 1992), speech-accompanying hand and arm gestures are primary candidates for extending the communicative capabilities of social robots. Not only are gestures frequently used by human speakers to illustrate what they express in speech (Cassell et al., 1998; McNeill, 2005), more crucially, they help to convey information which speech alone sometimes cannot provide, as in referential, spatial or iconic information (Hostetter, 2011). At the same time, human listeners have been shown to be well-attentive to information conveyed via such non-verbal behaviors (Goldin-Meadow, 1999; Hostetter, 2011). In addition, providing multiple modalities helps to dissolve ambiguity that is typical of unimodal communication and, consequently, to increase robustness of communication. Therefore it appears reasonable to equip humanoid robots that are intended to engage in natural and comprehensible HRI with co-verbal gestures.

Although gesture may be extended to include gaze, head and eye gestures, facial expressions, and body movements that manipulate objects in the environment, in this thesis the term *gesture* is used to refer specifically to hand and arm movements with a communicative intent. That means, gestures convey conceptual information which distinguishes them from other – arbitrary or functional – motor movements performed by the robot, hence the thesis title "conceptual motorics". The motivation to particularly focus on robot gesture in the present work is manifold and includes the following aspects.

- At the present time, only few scientific approaches have addressed the design and generation of social hand gestures for robots, and in effect, even fewer empirically validated design guidelines exist. A common practice in the generation of robot gesture is for roboticists to design and model non-verbal behaviors in a way that they consider suitable and realistic, without necessarily building upon theories from gesture research.

- In order to develop robots with human-like behaviors, researchers need to understand these behaviors in detail, for example, how such body movements are generated, how they can be related to conveying communicative intent, and what contextual characteristics lead to a better understanding of gestures. For this, social robots should be used as platforms for research on human communication, anthropomorphism, and embodiment. As a result, modeling human behavior on robots does not only help to equip them with enhanced social capabilities, but it also advances our understanding of how such mechanisms function in humans.

• Previous research in social HRI that has addressed the issue of gesture generation has not sufficiently and systematically evaluated people's perception and understanding of such non-verbal behaviors in robots with human-like embodiment. However, investigating the effect of robot gesture is considered key to improving gesture-based HRI and a step toward more sociable and acceptable robots in the future.

1.2 Objectives and Contributions

The work presented in this thesis aims to systematically address the above described challenges with a humanoid robot in an interdisciplinary approach. The following two main objectives combine both technically and psychologically inspired research and lead to the outlined contributions in each field:

(1) The first – technically motivated – objective is to develop and implement a robot control architecture for 'conceptual motorics', i.e., meaningful hand and arm movements that convey communicative intent and which can thus be considered conceptual. The resulting framework should enable a humanoid robot to flexibly produce synchronized speech and gesture at run-time, while not being limited to a predefined repertoire of motor actions. This requires an interface that combines conceptual representations of meaning and communicative intent as occurring in dialogue with a motor control layer for speech and hand movements. The chosen approach draws upon experiences already gained with the development of the Articulated Communicator Engine (ACE; Kopp and Wachsmuth, 2004), which represents the speech and gesture production model underlying the virtual agent *MAX* (see **Figure 1.2**). Given the different constraints encountered in the domain of physical robots, the implemented system should further extend the original multimodal scheduler in ACE with additional features, namely a predictive forward model and

4

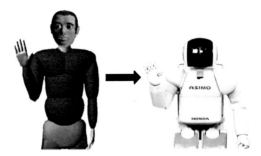

Figure 1.2: The goal of the present work is (1) to realize speech and non-verbal behavior generation for a physical humanoid robot (right) by transferring an existing virtual agent framework as employed for the conversational agent *MAX* (left), and (2) to subsequently evaluate it in controlled experiments of human-robot interaction.

an interactive feedback mechanism, to achieve optimal synchronization of robot gesture and speech. The contribution of this first major objective will thus be the realization of a multimodal action generation framework that is specifically tailored to the requirements of speech and gesture synthesis for the humanoid robot. Ideally, on a more abstract level, providing a valid solution for this specific robotic platform will represent a proof of concept that will demonstrate the feasibility of the chosen approach, namely employing a virtual agent framework for behavior realization in arbitrary physical humanoid robots. Accordingly, in future it should be possible to transfer the developed system to any other robotic platform with humanoid embodiment.

(2) The second – psychologically motivated – objective is to exploit the achieved flexibility in robot gesture, based on the implementation of a robot control architecture for 'conceptual motorics', for controlled experimental gesture studies. Since only very few studies in the area of HRI have so far focused on the perception of robotic hand gestures, this research objective will complement existing work by providing the field of social robotics with new empirical findings. On the micro-level, it should be investigated how humans perceive and understand gestural patterns performed by the humanoid robot. For this purpose, functions and effects of various speech-gesture deliveries should be examined in an interactional context. A macro-level analysis, on the other

hand, is intended to reveal how people experience the robot in dependence on its non-verbal communicative behavior during interaction. It should further explore whether and how the use of gesture affects the mental models humans form of a humanoid robot during interaction. The findings of these experimental studies aim to advance HRI research by giving new insights into human perception and understanding of social non-verbal behaviors in robotic agents. Consequently, the second major objective will contribute towards novel approaches in designing and building better artificial communicators.

1.3 Research Questions

The two major objectives stated above have been broken down into a set of research questions. A research question (RQ) is broader and less stringent than a hypothesis by formally stating the goal of the research. The following list summarizes the central research questions that have substantially driven the present work and are addressed in various parts of this thesis.

RQ1: Why should humanoid robots display non-verbal behavior such as hand gesture? ($\rightarrow$ Sections 1.1 and 2.2.2)

RQ2: What distinguishes the present approach from other existing solutions? ($\rightarrow$ Sections 3.2.1 and 3.3)

RQ3: What are the main challenges that have to be tackled when transferring a speech-gesture synthesis framework (i.e., ACE) from a virtual agent to a physical humanoid robot? ($\rightarrow$ Section 4.3)

RQ4: What are possible approaches to the realization of an ACE-based robot control architecture? Which approach represents the most suitable solution and why? ($\rightarrow$ Sections 5.1 and 5.2)

RQ5: Given the physical constraints of the humanoid robot (e.g., velocity limits, variation in mechanical degrees of freedom), how can the synchronization of robot gesture and speech be optimized to account for these limitations? ($\rightarrow$ Sections 6.1.3, 6.2, and 6.3)

RQ6: How do human interaction partners accept and evaluate multimodal robot behavior generated with the implemented framework? More specifically, how does non-verbal behavior, in particular hand gestures, of a humanoid robot impact and shape the human's interaction experience and assessment of the robot, e.g., in terms of subjective likability? ($\rightarrow$ Sections 7.3 and 8.3)

RQ7: How does human-like behavior in a humanoid robot affect the anthropomorphic perceptions and mental models humans form of the robot during interaction? ($\rightarrow$ Section 8.3)

1.4   Scope of the Thesis

The focus of the present work lies on the generation and evaluation of multimodal robot behavior with a special emphasis on co-verbal hand gesture. In general, computational approaches to synthesizing multimodal behavior can be modeled as three consecutive tasks as displayed in **Figure 1.3** (adapted from Reiter and Dale, 2000): firstly, determining *what* to convey (i.e., content planning); secondly, determining *how* to convey it (i.e., behavior planning); finally, conveying it (i.e., behavior realization).

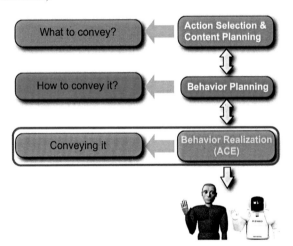

Figure 1.3: Behavior generation pipeline adapted from Reiter and Dale (2000); the work of this thesis focuses on the behavior realization at the lowest level of the pipeline.

By addressing the third task of this behavior generation pipeline, ACE operates at the behavior realization layer, yet the overall system used by the virtual agent MAX also provides an integrated content planning and behavior planning framework. The scope of this thesis is limited to ACE-based behavior realization at the lowest level of the generation pipeline, which forms the starting point for an interface endowing the humanoid robot with multimodal behavior. This means, we assume that the selection of appropriate actions and respective utterances is handled by other instances of the 'cognitive architecture' that drives the robot's behavior. Since addressing this issue would demand its own research project (see, for example, Kopp et al., 2008), strategies for the selection and planning of action, content, and behavior fall outside the scope of the present work.

Furthermore, the work presented in this thesis builds upon previous work that has been the subject matter of several years of research and which is described in detail in another dissertation (Kopp, 2003). An in-depth summary of the full functionality and often complex intricacies of the original ACE framework would thus go beyond the scope of this thesis. For this reason, it was decided to refrain from a detailed description of the complete system, but instead, to only highlight those aspects of it that are of explicit relevance to the above research questions.

The work described in this thesis was accomplished in two major steps in accordance with the research objectives stated in Section 1.2. Although the technical implementation of a speech-gesture generation framework (*objective 1*) is a prerequisite for conducting the experimental studies (*objective 2*), it is important to note that the primary purpose of these studies is not to evaluate the technical framework or to benchmark its functionality by testing different implementations. Instead, the aim is to utilize the realized framework as a tool for investigating more general research questions regarding the acceptance and evaluation of robot gestures by human interaction partners.

Finally, the work and results presented in this thesis are directly applicable only to the robot used; however, a certain potential of abstraction and generalization, with regard to both the technical implementation and the empirical evaluation studies, can be expected. Note that, if not stated otherwise, throughout this thesis the terms "robot" and "humanoid robot" refer to the Honda humanoid robot which served as the research platform for the present research (see Figure 1.2).

1.5 Structure of the Thesis

This thesis is divided into three core parts based on the intended scientific direction and contribution:

- **Part I** covers **background** information essential to the technical and empirical work described in later sections. Specifically, in **Chapter 2** fundamental results from communication research in psychology are summarized, describing in detail what is currently known about speech and gesture in human multimodal communication. In **Chapter 3** an overview of the current state of the art is provided with emphasis on the generation and evaluation of speech and gesture for artificial communicators. Models both for virtual conversational agents and for social robots are presented.

- **Part II** provides a description of the **technical implementation** realized on the humanoid robot for speech-gesture generation and synchronization. Specifically, in **Chapter 4** a system overview is given in the context of already existing modules and the required extensions. Moreover, the main challenges faced when transferring an action generation framework from a virtual agent platform to a physical robot are discussed. In **Chapter 5** the robot control architecture conceptualizing the generation of robotic hand and arm movements for gesture synthesis based on the virtual agent framework ACE is described. In **Chapter 6** the synchronization of the generated robot gesture with speech as an additional constraining modality is explained. In this context, an extended multimodal scheduler for finer and more flexible synchronization of the two modalities is presented.

- **Part III** introduces the **empirical evaluation** of the developed framework for speech and gesture generation with the humanoid robot. In particular, the set-ups, hypotheses and procedures of the two conducted experimental studies are described in **Chapters 7** and **8** respectively. Results obtained from an analysis of the collected data from each study are presented and discussed.

Finally, **Chapter 9** concludes the thesis with a summary and discussion of its contributions, especially with regard to the field of socially interactive agents (both virtual and robotic). The last section of the chapter outlines the scope for future research direction, highlighting desirable extensions to the realized speech-gesture generation framework, as well as further research avenues for the evaluation of robot gestures. It is hoped that these avenues will be tackled in the future and that they will further consolidate the contributions presented.

Resulting Publications

The work presented in this thesis has resulted in the following peer-reviewed publications: Salem et al. (2009, 2010a,b,c, 2011a,b,c); Salem (2011); Salem et al. (2012); Wachsmuth and Salem (*to appear*).

Part I: Background

Chapter 2

Gesture and Speech in Human Communication

In light of the motivation, objectives, and research questions introduced in the previous chapter, it appears sensible to provide an overview of some fundamental concepts that have been presented in the field of gesture research. Despite it being a relatively young research area, an extensive body of work investigating the phenomenon of human gesture has already been presented in the literature (e.g., McNeill, 1992, 2000, 2005; Kendon, 1997, 2004; Goldin-Meadow, 2003; Hostetter, 2011). Providing a comprehensive review or meta-analysis of all significant concepts and findings would thus go beyond the scope of this thesis. For this reason, only literature of particular relevance to the present research project is reviewed and discussed in this chapter.

Specifically, in Section 2.1 terminology that is relevant to the work outlined in this thesis and basic notions from the field of gesture research are introduced. Moreover, various approaches to classifying different types of gestures are presented and the structure of gesture is outlined. In Section 2.2 the combination of speech and gesture is described with a focus on the synchronization of the two modalities, as well as their semantic relationship. Note that speech-related findings presented in this section generalize only to stress-timed languages such as English, German, and Dutch; however, they may also apply to other languages. Finally, the purpose and use of gestures in multimodal communication is elucidated on the basis of different theories of gesture production models.

2.1 Gesture

Gesture is a phenomenon of human communication that has been studied by researchers from various disciplines for many years. A multiplicity of hand, arm and body movements can all be considered to be gestures, and although no universal or generally accepted definition of the term *gesture* exists, a great variety of definitions that emphasize different communicative aspects can be found in the literature. A selection of some definitions and classifications of gesture is presented in the following.

2.1.1 Definition of Gesture

According to Oxford Dictionaries Online, the word *gesture*, when used as a noun, refers to "a movement of part of the body, especially a hand or the head, to express an idea or meaning"[1]. Although accounting for the communicative quality of this type of non-verbal behavior, this definition does not provide information about the relationship between such body movement and spoken language.

In contrast, in his early work, Kendon (1980) refers to "speech-associated hand and arm movements [which are] to be distinguished from other kinds of bodily movement that can be observed in interaction" (p. 207) as 'gesticulation'. Kendon (2004) further mentions that "for an action to be treated as 'gesture' it must have features that make it stand out as such" (p. 10), i.e., listeners can identify a gesture as part of the meaning of the spoken utterance (Kendon, 1997). Unlike task-oriented movements like reaching or object manipulation, the characteristic shape and dynamical properties of gestures enable humans to distinguish them from subsidiary movements and to perceive them as meaningful (Wachsmuth and Kopp, 2002). Based on this definition, accidental gestures, fidgeting and so-called 'self-adaptors' (Kendon, 1980), i.e., movements in which the speaker touches or manipulates his own body, are not considered gestures. McNeill and Levy (1982) support this discriminating view by defining gesture as "any visible movement of the hand(s) excluding selfadaptors (scratching the head, fixing the hair)" (p. 5). This implies that gestures differ from other body movements on the level of intention, namely the intention to communicate (see also Melinger and Levelt, 2004). Accordingly, Väänänen and Böhm (1993) define gestures as "body movements which are used to convey some information from one person to another". Importantly though, and in accordance with other gesture researchers (e.g., McNeill, 1992; Iverson and Goldin-Meadow, 1998; Beattie, 2003), Kendon (1980) states that gesture is an integral part of speaking, meaning that gesture

---

[1]http://oxforddictionaries.com/ – accessed September 2011

14

and language form an integrated communicative system. In fact, about 90 % of hand gestures are generated in association with speech (McNeill, 1992), and are predominantly found to be performed in a spontaneous and idiosyncratic (i.e., speaker-dependent) manner (Cassell, 1998).

In line with these findings and the above definitions, the present work focuses on hand gestures that are produced along with speech, i.e., so-called *co-verbal gestures*, and which moreover pursue a communicative intent. However, gesture researchers often assume that some types of gestures communicate more than others (Hostetter, 2011). Since the focus of this thesis is on *communicative* robot gesture, the present investigations are limited to those types of gestures that are believed to be particularly communicative. Along with a classification and discussion of other gesture types, the gesture categories central to this research project are reviewed and outlined in the following section.

2.1.2   Classification of Gesture

Although the focus of this thesis is on spontaneous co-verbal gestures, it is helpful to first place them into a broader context to promote a clear understanding of their distinctive properties. Much gesture research has sought to describe and categorize the different types of gesture (e.g., Efron, 1972; McNeill, 1992; Kendon, 2004), leading to a great variety of definitions and categorizations that are based on different distinction criteria.

One such categorization was originally proposed by Kendon (1988), which was later arranged along an axis and labeled by McNeill (1992) as *Kendon's Continuum*. It is illustrated in **Figure 2.1**. As one moves from left to right along this axis, i.e., from gesticulation to sign languages, three reciprocal changes are to be noted, which are further illustrated by the following three dimensions (based on McNeill, 1992, 2000, 2005).

1) **Relationship to speech.** The extent to which speech is an obligatory accompaniment of gesture *decreases*.

2) **Relationship to linguistic properties.** The extent to which gesture shows properties that are typical of language *increases*.

3) **Relationship to conventions.** The extent to which gesture follows agreed conventions *increases*.

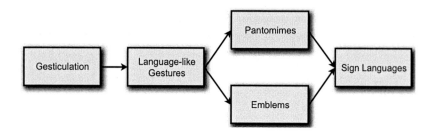

Figure 2.1: *Kendon's Continuum,* adapted from McNeill (1992, 2000).[2]

Gesticulation, based on the above definitions referred to as 'gesture' in this thesis, is obligatorily accompanied by speech. At the same time, its properties are the least language-like, i.e., such hand movements do not show the traits that are typical of a linguistic system. Finally, gesticulations are not conventionalized, since they are idiosyncratic and thus do not follow any specified rules.

Language-like gestures resemble gesticulations in form and appearance, however, they form parts of sentences by occupying a grammatical slot that replaces speech, e.g., "the meeting went [*gesture*]", with a gesture expressing the word 'so-so', for example by twisting the flat open hand. As a result, such gestures are performed in association with, but do not coincide with speech, and this way they take on some traits of a linguistic system. Similar to gesticulations, they are typically not conventionalized, since different gestures can be used to express the same word. In the given example with the word 'so-so', for example, a swinging head movement or sceptic facial expression could be used instead of the twisting hand gesture.

Pantomimes, by definition, are not accompanied by speech. In effect, while replacing spoken language, they show linguistic properties in that they can be combined into a sequence of gestures. To be correctly interpreted and understood by the observer, pantomimes follow some socially regulated conventions.

---

[2]In the original version from 1992, pantomimes were ordered before emblems, however, McNeill noted that their ordering was probably arbitrary. In a later version published in 2000, he uses both an ordering in which emblems appear before pantomimes (p. 2) and one in which pantomimes are listed before emblems (p. 3), depending on the relationship dimension considered. For this reason, it was decided to portray them on a coequal level in this figure.

Emblems such as the 'V-for-victory' sign or 'OK' sign have speech-independent symbolic status. They show limited language-like properties in that they must meet certain constraints of well-formedness in order to be interpreted correctly. For example, the thumb and the little finger cannot be used to form the 'OK' sign. However, they lack syntactic potential and compositionality, i.e., it is not possible to combine multiple emblems into a gesture sequence. Finally, emblematic gestures are locally or even globally conventionalized signs and are thus commonly understood within their cultural range.

Sign languages consist of socially regulated signs and are characterized by obligatorily absence of speech – in fact, simultaneous speaking (in non-deaf individuals) has a disruptive effect on both modalities (McNeill, 2000). Comprehensive linguistic structures that follow the same essential properties of all spoken languages, as well as the fully conventionalized nature of these signs within a given community ultimately justify the position of sign languages at the end of Kendon's Continuum.

Based on these observations, the relationship between the three reciprocal changes along Kendon's Continuum as listed above can be summarized as follows. Verbal and non-verbal aspects of human communication are complementary: spoken language is conventionalized, whereas gesticulation is idiosyncratic. Accordingly, the less the modality of speech is used, the more language-like and thus conventionalized the modality of gesture becomes in order to compensate for the lack of spoken language. Based on the three aforementioned dimensions, in comparison to the other points along Kendon's Continuum, gesticulation stands out with the following characteristics (McNeill, 2005):

- Only gesticulation is obligatorily accompanied by speech and thus integrated with linguistic content which, to the speaker who is unaware of his gesturing behavior, appears to be the main means of communication.

- Only gesticulation is unwitting, i.e., spontaneous, and not consciously produced like symbols or signs.

- Gesticulation does not follow any specified rules or conventions, although it is not unlikely to find similar gesturing patterns across different speakers.

Although gesticulation only forms one point on the Continuum, it represents the dominating gesture type (over 99 % of gesture) in conversation, discourse, storytelling, and spatial description (McNeill, 2005). Gesticulations are therefore also referred to as *conversational gestures* (Krauss et al., 1996). Since the focus of this thesis is on such speech-accompanying gesticulations, they are specified in more detail in the following.

Types of Co-Verbal Gestures

Over the past few decades, there have been several proposals to further classify the category of spontaneous co-verbal gesticulation into systems of gesture types, most of which were inspired by the early work of Efron (1941), Ekman and Friesen (1969), and Wundt (1973). These systems primarily vary in the number of gesture classes they suggest. One of the most accepted classification schemes in gesture research literature was introduced by McNeill (1992). Drawing inspiration from the semiotic categories suggested by Peirce (1960), he distinguishes four main types of co-verbal gestures: *beats, deictics, iconics,* and *metaphorics.* In his later work, however, McNeill (2005) claimed that the search for categories actually seems misled: since the majority of gestures are multifaceted, it is more appropriate to think in terms of combinable dimensions rather than categories. In this way, dimensions can be combined without the need for a hierarchy. For the sake of simplicity though, at this point we adopt the four gesture classes originally suggested; a more detailed description of each class is given below.

- **Beat gestures** are simple, repetitive, oscillating hand movements performed along with the rhythmical pulsation of accompanying speech without conveying any obvious semantic content (Feyereisen et al., 1988; Hostetter, 2011). They have been termed 'beats' because they resemble the beating of musical time as performed by an orchestra conductor.[3] Unlike other gesture types, beat gestures tend to have the same form regardless of the speech content (McNeill and Levy, 1982). Typically, they consist of only two movement phases in which simple flicks of the hands are performed either in an up- and downward or back- and forward direction. Beat gestures relate to the rhythmic structure of speech by synchronizing with its prosody, even though this relationship has been found to

---

[3]Originally, they have been named 'batons' in other classification schemes (Efron, 1941; Ekman and Friesen, 1969), hence referring to the instrument itself rather than the movement performed when using it. Other names suggested in the gesture literature include 'motor movements' (Hadar, 1989; Krauss et al., 1996), 'punctuating movements' (Freedman, 1977), and 'speech-marking hand movements' (Rimé and Schiaratura, 1991).

be more complex than often claimed in the literature (see McClave, 1994). By stressing the relevant parts of the discourse, gestural beats mark the information structure of the utterance. They further serve a pragmatic function when they accompany meta-comments on the speaker's own linguistic discourse and speech repairs (Cassell et al., 1998).

- **Deictic gestures** are classical pointing gestures that are typically performed with an extended index finger, "although any extensible object or body part can be used, including the head, nose, or chin, as well as manipulated artifacts" (McNeill, 1992, p. 80). Such gestures serve to indicate or spatialize discourse entities, e.g., locations, events, persons, objects, or object properties. These entities can either be concrete, i.e., physically present in the speaker's gesture space, or of an abstract nature, i.e., pointing at non-physical targets. An example for the use of an abstract deictic gesture is pointing to the side while saying "no, I meant [the other movie]". In conversations, discourse, and narratives, the majority of pointing gestures are of this abstract kind, while they are rarely if ever used to point at concrete entities (McNeill, 1992). While concrete pointing can be observed in young children before their first birthday, abstract pointing is typically only employed from the age of twelve (McNeill, 2005).

- **Iconic gestures** depict features of semantic content, which are also present in speech, by means of similarity between the form or manner of the gesture and its speech referent (McNeill, 2005; Ekman and Friesen, 1969). Such features of the semantic content can refer to concrete entities (e.g., triangular gesture depicting the shape or size of the 'give way' traffic sign), actions (e.g., circular gesture describing a rolling movement), or events (e.g., fist moving downwards to mime a falling rock). When describing spatial concepts, actions, or events, iconic gestures can reveal information about the speaker's viewpoint and imagistic mental representations of the described content (Alibali et al., 1999, 2000; Beattie and Shovelton, 2002; Alibali, 2005). These have been shown to differ across cultures and languages (Kita and Özyürek, 2003), as spatial and lexical information is conceptualized and processed differently depending on the speaker's cultural background (Kita, 2009). However, despite this tight connectedness and co-expressiveness of iconic gesture and speech, the information conveyed in each modality is not necessarily identical (Church and Goldin-Meadow, 1986; Roth, 2002; McNeill, 2005). This means, for example, that iconic gestures may describe the manner in which an action is carried out, even if this information is not given in the accompanying verbal expression (Cassell et al., 1998).

- **Metaphoric gestures** are similar to iconics in that they depict images, however, rather than representing concrete entities, they refer to abstract features, concepts, or ideas. Thus they have also been labeled 'ideographics' in other classification schemes (Efron, 1941; Ekman and Friesen, 1969). An example for a metaphoric gesture is a circular movement performed with the hand while saying "I told him over and over again". By representing an abstract feature of the described content, such gesture combines an iconic component with a metaphoric component (McNeill, 1992). In the given example, the iconic component consists of the circular form of the gesture, while at a metaphorical level, the circle expresses the notion of endlessness. Due to their more complex nature, it is not always easy to distinguish metaphoric from iconic gestures; this fact is also reflected in the more recent demand to think of combinable gesture dimensions rather than different types (McNeill, 2005; Krauss et al., 2000).

The latter two gesture classes, i.e., iconic and metaphoric gestures, are also referred to as *representational gestures* (Hostetter, 2011), since they particularly represent the content of speech by conveying or indicating meaning. Although McNeill's quadruple classification scheme is not considered to be accounting for all types of co-verbal gestures that can occur in conversational contexts, it still covers the majority of hand gestures that can be observed in narrative discourse. In fact, many researchers have adapted this four-way distinction, while partly modifying, simplifying, or adding further categories to it. In this regard so-called *interactive gestures* (Bavelas et al., 1992), i.e., movements that regulate interaction (especially turn-taking) between speakers, are worth mentioning.

A useful aggregation of McNeill's gesture classes was suggested by Krauss et al. (2000) who proposes two main categories of conversational gestures: *motor movements*, which are equivalent to beat gestures, and *lexical movements*, which are related to the semantic content of the accompanying speech and comprise McNeill's classes of deictic, iconic, and metaphoric gestures. The focus of the present work is on such lexical movements, and in the following of this thesis (if not stated otherwise), any occurrence of the term *gesture* will be referring specifically to representational (i.e., iconic and metaphoric) and deictic gestures. This is based on the assumption that these types of gesture are particularly communicative

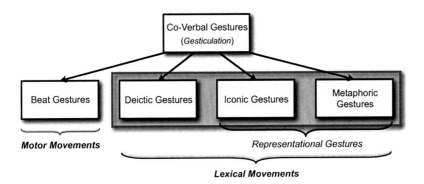

Figure 2.2: Classifcation of co-verbal gestures, adapted from McNeill (1992), Krauss et al. (2000), and Hostetter (2011); the gesture types that fall into the scope of the present work are framed in red.

(Hostetter, 2011). Other types such as beat gestures as well as interactive gestures fall outside the scope of the present work. **Figure 2.2** summarizes and illustrates the types of co-verbal gestures reviewed in this section and highlights the ones that are relevant to the remainder of this thesis.

2.1.3   Structure of Gesture

Kendon (1972) observed a hierarchical organization of human body motion which, in his later work (Kendon, 1980), he applied to specifically describe the structure of gesture. Based on his analysis of the hierarchy of gesture movements, Kendon distinguished between what he called a *gesture unit* ('G-Unit'), *gesture phrase* ('G-Phrase'), and *gesture phase*. Located at the top level of the hierarchy and thus forming the largest time interval, a **gesture unit** describes the period between two successive rest positions of the limbs. That means, it comprises all hand and arm movements that are performed from the moment the limbs start moving away from a rest position, e.g., arms hanging down, until the moment they return to one. It may contain one or several gesture phrases. A **gesture phrase** in turn is what is typically referred to as a 'gesture'. It can consist of up to five gesture phases, which are ordered in a sequence over time. Kendon initially differentiated among *preparation*, *stroke*, and *retraction* phase; Kita (1990) later added the two

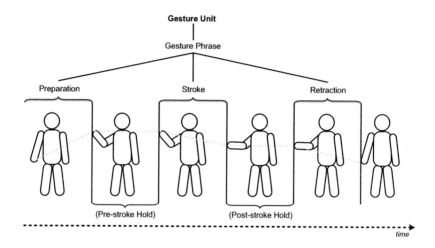# 2. GESTURE AND SPEECH IN HUMAN COMMUNICATION

Figure 2.3: Hierarchical gesture structure and temporal gesture phases as suggested by Kendon (1972, 1980) and Kita (1990).

non-obligatory *pre-stroke hold* and *post-stroke hold* phases. The hierarchical and temporal structure of gestures is illustrated in **Figure 2.3**; the different gesture phases are described in more detail below.

- **Preparation.** The hand moves away either from the rest position or from the end position of a previous gesture and is brought into the position in gesture space from which it can begin the upcoming stroke. The preparation phase typically anticipates the co-expressive linguistic segments which convey the gesture's meaning (McNeill, 1992). Kita et al. (1998) note that the preparation can potentially begin with a *liberating movement*, e.g., unfolding crossed arms, before the actual target-directed preparation movement starts. The latter can be further divided into two phases: *location preparation*, in which the hand is brought to the designated stroke start position, and *hand-internal preparation*, in which the shape and orientation of the hand are adjusted. Location preparation always precedes and typically overlaps with hand-internal preparation.

- **Pre-stroke hold.** Once the start position including the hand shape and orientation for the commencement of the stroke has been reached, this posture

22

might be briefly ceased in a pre-stroke hold. This is typically done to delay the stroke until the corresponding linguistic segment is ready to be expressed.

- **Stroke.** According to McNeill (2005), the stroke – also referred to as the *nucleus* of a gesture phrase (Kendon, 1972) – is the expressive phase of the gesture which bears meaning in relation to concurrent speech. It is often characterized as the most effortful gesture phase (Kendon, 1980; McNeill, 1992), in the sense that this phase consumes the most energy and exertion due to maximized acceleration. However, this phenomenological definition is problematic when two-phase gestures such as deictics are concerned, in which the preparation is followed by a motionless hold rather than an effortful stroke movement. For these cases, Duncan (quoted in McNeill, 2005) extended the gesture phase classification with the concept of *stroke hold* phases for non-dynamic strokes. This notion was further supported by Kita et al. (1998) who differentiated among two different types of holds: firstly, *independent holds* which equal stroke holds and are so termed as they can occur on their own; secondly, *dependent holds* which correlate to pre- and post-stroke holds and are so named due to their dependence on the gesture strokes.

- **Post-stroke hold.** Having reached the final position at the end of the stroke, the hand may freeze in this posture for a more or less brief moment before the retraction begins. Such post-stroke hold is typically performed if the corresponding linguistic segment has not been fully articulated yet, while the stroke itself has been completed. Similar to the pre-stroke hold, it thus serves the synchronization of the gesture stroke with accompanying speech. Moreover, it may extend and stress the meaning conveyed by the gesture stroke for the duration of the hold (Kita, 1990).

- **Retraction.** The hand subsides to a rest position which, however, does not necessarily have to be the same as at the beginning. Especially if the relaxation is temporary due to another upcoming gesture phrase, only a *partial retraction* (Kita et al., 1998) may be performed, before the actual rest position has been reached. Such anticipation of the upcoming gesture may cause the retraction phase of one gesture to become one with the preparation phase of the following gesture (Cassell et al., 1994), thus leading to a direct transition. From a functional point of view, the start of the retraction phase reveals "the moment at which the meaning of the gesture has been fully discharged" (McNeill, 2005,

p. 33). Other terms used to describe the retraction phase are *recovery* or *return* phase (Kendon, 1972, 1980).

Note that except for the stroke, all other gesture phases are optional and can be either partially or completely skipped, especially when multiple gesture phrases are present.

2.2 Combining Speech and Gesture

In view of the first major research objective of the present work, namely to enable a humanoid robot to produce synchronized speech and gesture, it is reasonable to model such a concept based on multimodal communication as observed in the human counterpart. The combination and synchronization of speech and gesture has been a central aspect of investigation among researchers from various disciplines including Psychology, Linguistics, and Neuroscience. A growing body of empirical evidence has led to the general view that gestures are a way of expression that is tightly linked to speech and language respectively (e.g., McNeill, 1992; Goldin-Meadow, 2003; Kendon, 2004).

The exact relationship between the two modalities, however, has not yet been entirely deciphered. According to Kendon (2007) the theoretical models trying to formalize this issue can be roughly divided into two different views: on the one hand, *"speech auxiliary* theories" (p. 7–8) regard speech as the primary modality and gesture as aids to the speaker (e.g., supported by Freedman, 1977; Rimé and Schiaratura, 1991; Kita, 2000; Krauss et al., 2000); on the other hand, *"partnership* theories" (p. 8) consider gesture and speech to be co-operative and equal partners in the production of multimodal utterances (e.g., supported by McNeill, 1992; Clark, 1996; Gullberg, 1998; de Ruiter, 2000; Kendon, 2004).

Despite the controversy, findings from human-related studies are being increasingly applied to research areas focusing on artificial communication, e.g., for human-machine interaction, computer animation or social robots. In order to model realistic and acceptable communicative behavior for an artificial communicator such as an embodied conversational agent (ECA) or a humanoid robot, a fine grasp of the precise interplay between speech and gesture is crucial. For this purpose, reference units for both modalities must be identified and precisely specified before putting them into a mutual context. Thus, the following subsections are aimed at elucidating empirical findings with regard to the relationship between speech and co-verbal gesture in human communication. Various perspectives as well as terminological definitions that are relevant to the technical implementation presented in the second part of this thesis are thereby discussed.

2.2.1   Speech-Gesture Synchrony

Following the above mentioned controversy regarding the general relationship between speech and gesture, i.e., whether one modality is superior to the other or whether they are operating at the same level, the issue of synchronization is equally inconclusive. As de Ruiter (2000) notes, synchronization is difficult to define, since the "synchrony between gesture and speech is the synchrony between two time intervals that are often hard to define" (p. 297–298). Nevertheless, a number of gesture researchers have attempted to define these intervals more precisely; a brief history of their work is summarized in the following.

Temporal Relationship between Speech and Gesture

In the early 1960s, Condon was one of the first to carry out thorough analyses to investigate the temporal relationship between speech and body movement. He observed that a speaker's movements are finely synchronized with his own speech, which led him to suspect a common neurological basis of the two modalities and to term their relationship *self-synchrony* (Condon, 1976).

Kendon (1972, 1980) built on Condon's findings and extended his work by specifying in more detail the hierarchical components of gesture units and defining their different gesture phases (see Figure 2.3). He identified the stroke to be the dynamic peak of the gestural movement, thus forming the smallest unit of a gesture. More importantly though with regard to the temporal relationship between speech and gesture, Kendon discovered that the gesture stroke generally begins shortly before or right at the onset of the stressed syllable in speech, but never follows. Note, however, that opposed to generalizations that had been previously made by Schegloff (1984) and McNeill (1992), he later found that the time of completion of the gesture stroke does not have to coincide with, but can indeed follow the tonic syllable of the co-occurrent speech (Kendon, 2004).

Kendon further described a correspondence between the different hierarchical levels in which both gesture and speech are organized. At a higher level, gesture units align with what he called *locutions*, which typically comprise complete sentences (Kendon, 1980). In the same way that gesture units can consist of one or more gesture phrases, locutions accordingly consist of a number of prosodic phrases, so-called *tone units* (based on the definition by Crystal and Davy, 1969). Similar to how gesture phrases represent meaningful units of bodily action, tone units are "packages of speech production identified by prosodic features which correspond to units of discourse meaning" (Kendon, 2004, p. 108), e.g., as given by sub-clauses in a sentence.[4] This concerted expression of meaning at the level of tone unit and gesture phrase is what Kendon (1980) termed *idea unit*, which led him to believe that speech and gesture originate from one and the same underlying process. Finally, at the lowest level of both hierarchies, the gesture stroke corresponds to the *nucleus* of the prosodic phrase, which is represented by the tonic (i.e., most prominent, stressed) syllable in speech. The hierarchical organization and synchronization of speech and gesture as suggested by Kendon (1972, 1980) is illustrated in **Figure 2.4**.

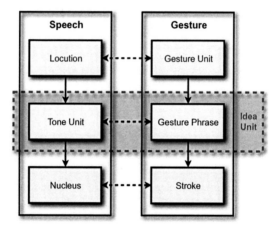

Figure 2.4: Hierarchical organization and synchronization of speech and gesture based on Kendon (1972, 1980).

---

[4]In view of this definition, tone units are similar to what Levelt (1989) specified as *intonation phrases*, which in turn consist of a series of *phonological phrases*, i.e., one or several connected words of spoken language.

Kendon (1980) notes, however, that below the locution and gesture unit level respectively, the association between entities of the two modalities is less stringent. For example, tone units do not necessarily correspond to gesture phrases on a one-to-one basis, i.e., one tone unit can be associated with more than one gesture phrase and vice versa, although this rather represents the exceptional case.

Partially building on Kendon's work and his idea of a common origin for both modalities, McNeill (1992) states that "gesture and speech have a constant relationship in time" (p. 25). This is based on the assumption that gestures generally anticipate and synchronize with speech, which seems sensible in light of the optional preparation as well as hold phases before and after the gesture stroke. To clarify the temporal relationship between speech and gesture, McNeill formalized three synchrony rules that apply specifically to the stroke phase of the gesture (p. 26–29):

- **Phonological Synchrony Rule.** Based on observations described by Kendon (1980), the gesture stroke precedes or ends at – but never follows – the "phonological peak syllable of speech" (McNeill, 1992, p. 26). This suggests an integration of the stroke phase into the phonology of the spoken utterance.

- **Semantic Synchrony Rule.** If speech and gesture occur together, the same meanings are conveyed by both modalities at the same time. That means, as gestures supplement or complement the content of speech, both channels express the same *idea unit* (McNeill, 1992). According to de Ruiter (2000), this type of synchrony "follows from the fact that both gesture and speech ultimately derive from the same communicative intention" (p. 304). The semantic synchrony rule applies even when speech is interrupted by pauses, or in cases with multiple gestures or clauses.

- **Pragmatic Synchrony Rule.** If speech and gesture occur together, they "perform the same pragmatic functions" (McNeill, 1992, p. 29). This rule is particularly applicable to metaphoric co-verbal gestures, which might not be directly related to the speech content on a strictly semantic level, but rather on a pragmatic level. For example, the spoken utterance "the meeting went on and on" accompanied by a circular gesture movement representing the lasting nature of the meeting depicts a case of co-expressivity on the pragmatic level. Pragmatic synchrony can often have an emphasizing or contrasting effect which highlights the relevant information of the utterance (Kendon, 2004).

2. GESTURE AND SPEECH IN HUMAN COMMUNICATION

Although McNeill's synchrony rules were originally derived from a limited number of direct observations, their validity has been backed up by several more recent empirical studies, for example, Loehr (2007), Özyürek et al. (2007), and Kelly et al. (2004).

Definition of Reference Units of Speech-Gesture Synchronization

Despite the general acceptance of above described synchrony rules, researchers such as de Ruiter (2000) point out that the definition of synchronization is more complex and problematic than it may seem:

> "*The conceptual representation (the 'state of affairs' [Levelt 1989] or the 'Idea Unit' [McNeill 1992]) from which the gesture is derived might be overtly realized in speech as a (possibly complex) phrase, and is not necessarily realized overtly as a single word. Therefore, it is by no means straightforward to unambiguously identify the affiliate of a given gesture.*" (p. 297)

McNeill (2005) agrees that the identification of what the gesture literature often refers to as the *lexical affiliate* can indeed prove difficult, since it has been variously interpreted. He adopts the definition introduced by Schegloff (1984) by specifying the lexical affiliate to be the word(s) corresponding most closely to a gesture in terms of meaning. This affiliate, however, has to be distinguished from the co-expressive speech segment that might synchronize temporally with a gesture, but not necessarily semantically. McNeill cites an explicative example of the distinction between the lexical affiliate and co-expressive speech originally presented by Engle (2000): a study participant attempting to describe a lock-and-key mechanism said "lift them to a height, to the perfect height, where it [**enables**] the key to move". The word "enables" was temporally synchronized with the stroke of an iconic gesture miming a key turning movement, thus representing the co-expressive speech segment. The lexical affiliate, namely "key" or "key to move", however, only occurred after the gesture stroke. McNeill attempts to explain such co-expressivity by interpreting the word "enables" combined with the turning gesture to convey the concept of 'being able to turn the key' by lifting them (i.e., the tumblers) up. This way, the newsworthy content is highlighted by means of complimentary, but temporally synchronized, information via both modalities. On another note, this

28

example and interpretation is in line with aforementioned claim stating that the gesture stroke precedes and thus anticipates the lexical affiliate.

To simplify matters, and in light of the technical objective of the present work, we shall define the term *affiliate* in the sense of a 'conceptual affiliate' rather than a 'lexical affiliate' as follows:

Definition 1: *The **affiliate** is defined as the co-expressive word or sub-phrase in speech that conveys the same idea unit as its corresponding gesture.*

To achieve this co-expressive synchrony, the gesture stroke onset generally precedes or, at the latest, begins right at the onset of the stressed syllable (i.e., the nucleus) of the affiliate. In the case of a static stroke (e.g., stroke hold of a deictic gesture), the stroke phase is assumed to last for the duration of the affiliate; in the case of a dynamic stroke (e.g., of an iconic gesture), the stroke phase is assumed to be optionally followed by a post-stroke hold for the remaining duration of the affiliate. This definition is in line with McNeill's synchrony rules as described above and further follows Kendon's specification of a *tone unit*, which co-expresses an idea unit together with its associated gesture phrase (see Figure 2.4).

At a higher level of speech-gesture communication, we can define a *multimodal utterance* to represent an utterance (typically comprising one spoken sentence) in which at least one sub-phrase is accompanied by a gesture. Such multimodal utterance, in turn, is composed of one or multiple so-called *chunks*, which can consist of speech or gesture, or both. For the latter case, in which both modalities are used, we can adopt the following definition of a *multimodal chunk* from Kopp (2003, p. 21):

Definition 2: *A **multimodal chunk** of speech-gesture production is defined as a pair of an intonation phrase and a co-expressive gesture phrase. Within each chunk, the prominent concept (idea unit) is concertedly conveyed by a gesture and an affiliate.*

According to this definition, a multimodal utterance comprising several gestures is considered to consist of multiple chunks, since each chunk can only contain a single gesture. The hierarchical structure and organization of a multimodal utterance, as well as the synchronization within a multimodal chunk as described above is illustrated in **Figure 2.5**. Corresponding cross-modal synchronization

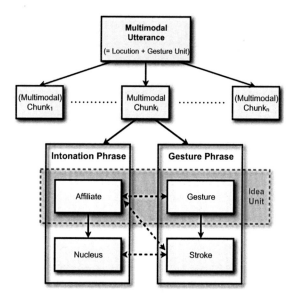

Figure 2.5: Hierarchical structure of a multimodal utterance and synchronization within a multimodal chunk; partly derived from Kendon (1980) and Kopp (2003).

dependencies between units of an intonation phrase and a gesture phrase are visualized by double-headed arrows.

Cross-Modal Adaptation

In order to model multimodal communication based on speech and gesture for an artificial communicator like a robot, it is important to understand how the two modalities adapt to each other during the process of synchronization. A common view, also expressed by Kopp (2003, p. 22) with reference to Levelt et al. (1985) and de Ruiter (1998), assumes that in multimodal communication, gesture predominantly and more flexibly adjusts to speech, especially following the onset of speech and gesture production. This type of alignment is typically achieved via hold phases which may be inserted before or after a gesture stroke. Such view is in line with what McNeill (1992) refers to as a gesture's anticipation of speech.

Kendon (2004), however, opposes this view by giving a series of examples which clearly demonstrate the existence of two cases: firstly, instances in which gesture adjusts to speech by means of adequately timed preparation or additional hold phases; secondly, instances in which speech adapts to gesture by means of pauses which do not arise from word searches or difficulties in speaking.[5] He summarizes his findings as follows:

> "*The precise way in which a coincidence is achieved between a gesture phrase and that part of the tone unit to which it is related semantically appears to be variable. In our interpretation, this means that the speaker can adjust both speech and gesture one to another as if they are two separate expressive resources which can be deployed, each in relation to the other, in different ways according to how the utterance is being fashioned.*" (p. 126)

In view of the technical objective of the present work and the need to model the synchronization process between speech and gesture for artificial communicators, we shall adopt the latter notion presented by Kendon. Thus, we assume cross-modal adaptation to be possible both for gesture by means of anticipating preparation and hold phases and for speech by means of pauses. However, on the basis of the findings presented by Levelt et al. (1985) and de Ruiter (1998), we further assume gesture adaptation to speech to predominate in speech-gesture synchronization. That means, synchronization between the two modalities is primarily achieved by means of gesture adapting to running speech, and secondarily by means of speech pausing to adapt to continuous gesturing.

A more detailed description and review of the relevant literature concerning the concept of speech-gesture synchronization can be found in de Ruiter (1998, p. 5–48) and Kopp (2003, p. 22–28, in German).

2.2.2 Semantic Relationship

Although gesture and speech occur in close temporal synchrony and typically convey the same *idea unit*, they can express their meanings in very different

---

[5]Although not explicitly mentioned by Kendon (2004), this implies an adaptation of the speech rate to gesture: generally, speech rate may be either decreased by means of pauses or by speaking more slowly, or it may be increased by speaking faster.

ways (McNeill, 1992). In other words, the semantic relationship between the two communication channels can vary depending on how the information conveyed by one modality relates to the other. Generally, the relationship between gesture and speech content can be divided into two broad classes which are described in the following.

1) **Redundant.** If a gesture conveys identical information as the accompanying speech, it is referred to as being *redundant* or *matching* (Hostetter and Alibali, 2008; Church and Goldin-Meadow, 1986). Using such gesture is considered to help illustrate or emphasize what is being said. For example, a speaker could draw a circle in the air with his index finger while saying "he took the [**round**] cup coaster". This iconic gesture is redundant in that it illustrates the object property, namely its round shape, when this attribute is simultaneously expressed via speech. By definition, the information conveyed in a redundant gesture is always congruent with the information content expressed via speech.

2) **Non-redundant.** If a gesture conveys information that is not simultaneously conveyed in the accompanying speech, it is referred to as *non-redundant, complementary, supplementary,* or *mismatching* (Alibali et al., 2000; Melinger and Kita, 2004; Hostetter, 2011; Church and Goldin-Meadow, 1986). The use of such gesture provides additional information which might also be important to the message conveyed, but is not explicitly expressed in speech. Like in the above example, a speaker could draw a circle in the air with his index finger, however, this time saying "he took the [**cup coaster**]". In this case, information about the shape of the object is only conveyed in the gesture. Moreover, non-redundant iconic gestures are frequently used to provide additional information about the manner in which an action or movement is performed, when this is not expressed in accompanying speech. This is particularly true for verbs that describe movement without specifying the exact manner of motion, such as "go", "come", or "leave" (Cassell et al., 1998). In some instances, gestures may communicate information that is complementary at the temporal or prosodical level of the speech affiliate, but not necessarily at the level of the discourse, e.g., if the same related information is given in a preceding or following phrase of speech (Alibali et al., 2000). Non-redundant gestures can be further subdivided into **congruent** and **incongruent** gestures in relation to speech. While gestures of the first subgroup convey information that is complementary but semantically consistent with what is expressed in speech,

the latter subgroup describes gestures whose information content contradicts the message conveyed in speech. However, except by mistake (e.g., when saying "turn right here" while pointing to the left side), such incongruent speech-gesture mismatches do not typically occur in the discourse of normal adults (Cassell et al., 1998). Yet they are occasionally used as stimuli in gesture studies, in which the semantic or temporal relationship between speech and gesture is deliberatively manipulated to investigate the resulting effects on the listener (e.g., McNeill et al., 1994; Cassell et al., 1998; Kelly et al., 2004; Özyürek et al., 2007; Galati and Samuel, 2011; Habets et al., 2011).

Given these two classes of possible semantic relationships between speech and gesture, the question regarding their potentially different communicative value arises. In other words, although both redundant and non-redundant gestures are likely to have an effect on communication, they may do so in different ways.

Addressing this issue as part of an extensive meta-analysis, Hostetter (2011) investigated whether gestures are more communicative – in terms of being more informative – when they are non-redundant with speech as opposed to when they are redundant. Based on a comparison of the findings from a number of studies, she comes to the conclusion that indeed gestures have a larger effect on communication when they are not completely redundant with the accompanying speech. This insight may not be surprising, since listeners arguably gain only little from seeing the speaker's gesture when all the important information is also expressed in speech.

Importantly though, research suggests that listeners generally notice information that is exclusively conveyed in a speaker's gesture and frequently integrate this information when later on asked to retell the content they had listened to (e.g., McNeill et al., 1994; Cassell et al., 1998). This implies that listeners frequently pick up the meanings expressed by gestures and incorporate them into their spoken narrative. Such 'speech-gesture binding', as McNeill (2005) calls it, has been shown to work both ways: Kelly et al. (1999) found that study participants recalled information that was actually expressed in speech as having been gestural. Accordingly, these experiments reveal that gesture and speech exchange semantic information freely and spontaneously in the listener's memory.

Despite the fact that non-redundant gestures communicate more than redundant gestures, both types of gesture have been shown to be beneficial to communication in multiple ways. Firstly, listeners who see a speaker's gestures have better immediate comprehension of the speaker's semantic intent than when speech is not accompanied by visible gestures (Hostetter, 2011, based on the computed mean effect size across 63 study samples in her meta-analysis). Secondly, several studies suggest that seeing gestures during interaction can be a beneficial aid for subsequent information retrieval and memory with regard to the communicated information (e.g., Kelly et al., 1999; Galati and Samuel, 2011).

These findings are particularly true for deictic and representational gesture (Kelly et al., 1999, 2004; Beattie and Shovelton, 2001; Feyereisen, 2006). Thus, they motivate the objectives of the present work, namely to endow a humanoid robot with these types of co-verbal communicative gesture and, furthermore, to investigate whether findings from human-human communication also hold true for human-robot interaction (see RQ1).

2.2.3 Function of Co-Verbal Gesture

The previous section has focused primarily on how a speaker's combined use of speech and gesture affects the listener in an interaction. In fact, traditionally, gestures were viewed to serve a **communicative function** by providing information to listeners about the semantic content of the speaker's utterance (Kendon, 1994; Krauss et al., 2000). Nowadays, however, this one-sided view is considered outdated, as it fails to account for the many complex ways in which gestures can contribute to communication, particularly in dependence on the type of gesture employed. But if gestures are not primarily produced to facilitate comprehension on the listener's side, what other function do they serve?

As a matter of fact, some researchers claim that gestures convey only little or no information to addressees, suggesting that gestures are not at all intended to communicate (Krauss et al., 1991). This view typically stems from observations disclosing that people gesture even when they cannot see each other (e.g., while speaking on the telephone), thus showing that comprehension on the listener's side and communication in general are still effectively possible without gesture. Rimé (1982), for example, measured the gesture rate in pairs of study participants who had been instructed to converse about movies either with or without a partition between them. He found that the frequency of gesture use was only slightly reduced when participants could not see each other compared to when they could. Furthermore, even blind speakers who have never observed gestures by others have been shown to gesture while they speak (Iverson and Goldin-Meadow, 1997).

Such findings have led some researches to believe that speakers do not use gestures for the purpose of communicating to the listener, but primarily for themselves, specifically to support lexical access in spontaneous speech (Rimé and Schiaratura, 1991; Krauss et al., 1991, 1996; Hadar and Butterworth, 1997). This view has been backed up by studies in which speakers were prevented from gesturing: compared to when they were able to gesture, the restriction was shown to decrease speech fluency with regard to spatial content (Rauscher et al., 1996) and to increase retrieval failures in a 'tip-of-the-tongue' situation (Frick-Horbury and Guttentag, 1998).

Given these findings, researchers increasingly share a view that attributes a **cognitive function** to gesture. This, however, does not necessarily preclude the communicative function traditionally assigned to gesture, especially since other studies showed that listeners do attend to information conveyed by the speaker's gestures, particularly when they are non-redundant with speech (Goldin-Meadow et al., 1992; McNeill et al., 1994). Moreover, there is evidence that, depending on the experimental task, speakers actually gesture more when the interlocutor can see them. For example, Cohen and Harrison (1973) and Cohen (1977) found that when asking participants to give route directions, far more (mainly deictic) gestures were used when speaker and listener were mutually visible than when they were not.

Bavelas and colleagues conducted similar experiments in which they manipulated the presence or visibility of interlocutors (Bavelas et al., 1992; Bavelas, 1994); the results led them to distinguish between *interactive gestures* and *topic gestures*. The purpose of interactive gestures is to help maintain the conversation, e.g., by cross-referencing new content to the general theme of the conversation, indicating agreement with or understanding of the other's contribution, or by managing turn taking. Hence, interactive gestures refer directly to the interlocutor and provide no information on the topic of discourse. In contrast, topic gestures depict aspects of utterance content, expressing semantic information that is directly related to the topic of conversation. Importantly though, Bavelas et al. showed that when speakers could not see each other, they refrained from using interactive gestures but continued to use topic gestures.

In a later experiment in which participants described a complex picture either in a face-to-face conversation, via telephone or in a monologue to a tape recorder, Bavelas et al. (2008) identified yet another dimension that influenced the speaker's use of gestures. They showed that not only visibility affects the type and frequency of gesturing – e.g., interactive and non-redundant gestures were most used in the face-to-face condition – but also highlighted the role of dialogue when speaking. They basically argued that, despite the lack of visibility, speakers in the telephone condition did not use significantly less topic gestures than in the face-to-face condition because they were engaging in a dialogue.

Furthermore, Bavelas et al. (2008) made a good case for the importance of the stimulus used in studies of this kind, arguing that findings from previous studies differed substantially due to the varying nature of experimental tasks. These ranged widely, e.g., from completely non-visual material (Krauss et al., 1995, Exps. 1 & 2, sound and taste stimuli), to abstract topics like discussing opinions on movies (Rimé, 1982), to giving spatial directions (Cohen and Harrison, 1973; Cohen, 1977), to describing a scene from a cartoon (Bavelas et al., 1992; McNeill et al., 1994), or other highly visual material (Bavelas et al., 2008). In fact, it was shown that the experimental stimulus characteristics, i.e., whether the assigned topic was likely to elicit a high gesturing rate or not, have a significant effect on measured gesture rates (Krauss et al., 1995; Bavelas et al., 2002).

In summary, past research on the function of gesture suggests that different types of gestures are used in different communicative settings to serve different – communicative or cognitive – functions. Depending on their function, in turn, gestures are likely to have different origins within the speaker's mind (Krauss et al., 2000). Indeed, many attempts have been made to conceptualize the exact origin of and mechanisms behind co-verbal gesture production, yielding a great variety of theoretical models to be found in the gesture literature. Since a number of extensive overviews presenting the various models of gesture production have already been given by other authors (e.g., Kopp, 2003; Bergmann, 2011), the following subsection will be limited to a brief summary of these different views.

Theoretical Models of Gesture Production

As becomes obvious from the above written, the exact function and thus the underlying model of gesture production in human discourse is still a subject to controversial discussion. The controversy is related to the two different views regarding the relationship between speech and gesture as mentioned at the beginning of Section 2.2: while one suggests that the two modalities are arranged in a functional hierarchy, typically with speech being the primary modality (Krauss et al., 2000), the second view considers speech and gesture to be operating as equal partners in communication (Kendon, 2007). This crucial distinction is reflected in the different psychological and psycholinguistic models and theories that have been proposed in the gesture literature, based on which they can be roughly divided into two sets (Gullberg et al., 2008).

The first set, which is based on the 'gesture as auxiliary' assumption, either views gestures as serving and facilitating lexical retrieval (*Lexical Retrieval Hypothesis*, Krauss et al., 2000) or considers gestures as contributing to the conceptual planning and packaging of imagistic content for verbalisation (*Information Packaging Hypothesis*, Freedman, 1977; Alibali et al., 2000). The second set of model-based theories regards gestures as an integral part of utterance production; however, they vary in focus. Either they support the 'gestures as a window into the speaker's mind' view (*Growth Point Theory*, McNeill, 1992, 2005; McNeill and Duncan, 2000), or they concentrate on the interaction between imagistic and linguistic thinking (*Interface Hypothesis*, Kita and Özyürek, 2003), or they focus on the communicative intention which drives the two communication channels to form a coherent multimodal utterance (e.g., *Sketch Model*, de Ruiter, 2000).

An illustrative view of the classification of the models presented in this section is displayed in **Figure 2.6**. Note that the list of theories and models respectively given in this short overview is by far not exhaustive, as new models are constantly emerging and old models are frequently being updated. For a more detailed review see, for example, Bergmann (2011).

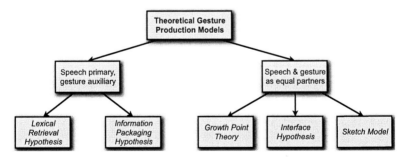

Figure 2.6: Theoretical models of gesture production; adaptation of classification scheme as proposed by Gullberg et al. (2008).

2.3   Summary

In view of the research objectives of this thesis, namely to endow a humanoid robot with synchronized speech and gesture and further to evaluate the robot's communicative behavior, it is reasonable to consider and investigate such multimodal communication in humans as a starting point. In fact, in order to model realistic and acceptable communicative behavior for an artificial communicator, a thorough understanding of the concept of speech-accompanying gesture as well as the precise interplay between the two modalities is crucial. For this reason, the relevant literature from the field of gesture research was reviewed in this chapter, focusing on psychological and psycholinguistic communication theories. Furthermore, definitions of relevant terminology were provided.

Initially, **Section 2.1** provided an isolated view on the phenomenon of human gesture. A number of definitions of the term *gesture* were first given in 2.1.1, before presenting *Kendon's Continuum* as a representative classification scheme for categorizing different types of gesture in 2.1.2. It comprises five different gesture classes ordered along a horizontal axis, namely (1) gesticulation, (2) language-like gestures, (3) pantomimes, (4) emblems, and (5) sign languages. Moving from gesticulation to sign languages along this axis, each gesture class varies with regards to its relationship to a) speech, b) linguistic properties, and c) conventions.

The first class, **gesticulation**, was identified to stand out with a number of characteristics: first, only gesticulation is obligatorily accompanied by speech; second, gesticulation is produced spontaneously and unwittingly; finally, it does not follow any specified rules or conventions. Representing the dominating gesture class in human discourse and conversation, gesticulation is also referred to as *conversational gesture* and was thus highlighted as being central to the work presented in this thesis.

Based on a subsequent, more detailed breakdown of the types of co-verbal gestures which the gesticulation class can comprise, three gesture types were pointed out as being relevant to the following parts of the thesis: (1) **iconics**, (2) **metaphorics**, and (3) **deictics**. The first two types taken together are also referred to as *representational gestures*; all three types together, in turn, have been labeled *lexical movements*, since they are related to the semantic content of the accompanying speech. From this point on, these three types will represent what is meant by the term **gesture**.

In Section 2.1.3, the structure of gesture was investigated, yielding a hierarchical organization on three levels. At the top level, a **gesture unit** spans the period between two successive rest positions of the limbs. It comprises at least one **gesture phrase**, which equals what is typically referred to as a 'gesture'. This, in turn, consists of up to five **gesture phrases**: *preparation, pre-stroke hold, stroke, post-stroke hold*, and *retraction*. Except for the stroke phase, all other gesture phases are optional and may be either partially or completely skipped, e.g., when multiple gestures are performed in a row.

Following the detailed description of the basic concepts related to gesture, in **Section 2.2** the combination of speech and gesture was brought into focus. Although no universal definition of the exact relationship between the two modalities exists, a number of aspects that shape this relationship were elucidated. At first, in 2.2.1, speech-gesture synchrony was reviewed with a focus on the temporal relationship that regulates the interplay between the two modalities. In particular, three temporal synchrony rules as proposed by McNeill (1992) were presented: (1) the **phonological synchrony rule** stating that the gesture stroke precedes or ends at, but never follows, the prominent syllable of speech; (2) the **semantic synchrony rule** claiming that speech and gesture convey the same meanings whenever they co-occur; (3) the **pragmatic synchrony rule** stating that speech and gesture serve the same pragmatic functions.

Subsequently, in anticipation of the technical implementation presented in the second part of this thesis, a couple of definitions were given to specify distinct reference units of the speech-gesture synchronization process: Definition 1 introduced the term **affiliate** which is specified as the word or sub-phrase that is associated with the accompanying gesture; Definition 2 specified the term **multimodal chunk** as a pair of an intonation phrase and a co-expressive gesture phrase which concertedly convey the common *idea unit*.

A further point of interest elucidated with regard to the synchronization of speech and gesture was the process of cross-modal adaptation. It was found that adaptation is possible both for gesture by means of anticipating preparation and hold phases, and for speech by means of pauses. It is assumed that synchronization is primarily achieved by means of gesture adapting to running speech, and secondarily by means of speech pausing to adapt to continuous gesturing.

Section 2.2.2 was dedicated to the investigation of the semantic relationship that speech and gesture may establish during the communicative process. Importantly, two relationship classes were identified: firstly, **redundant** gestures convey identical information as the accompanying speech; secondly, **non-redundant** gestures convey information that is not simultaneously conveyed in the accompanying speech. The latter class can be further subdivided into **congruent** and **incongruent** gesture, depending on whether or not the information conveyed in the gesture is semantically consistent with what is expressed in speech. With reference to empirical findings from the gesture literature, it was further found that gestures are more communicative when they are non-redundant rather than redundant with speech. Generally though, both types of gesture have been shown to be beneficial to communication, as they can foster both the listener's immediate comprehension and subsequent information retrieval and memory. These findings are particularly true for deictic and representational (i.e., iconic and metaphorical) gestures; thus, they substantiate the choice to focus on these gesture types in light of the objectives of the present work.

Finally, in Section 2.2.3, the function of co-verbal gesture was examined, yielding two major functions which gestures are viewed to serve: first, by providing information for the listener about the semantic content of the speaker's utterance, a gesture may serve a **communicative function**; second, as gestures have been shown to support the speaker in terms of lexical access in spontaneous speech, they are considered to further serve a **cognitive function**.

Concluding the review of this chapter, an overview of different theoretical models of gesture production was given. The theories and models respectively can be roughly divided into two sets: theories of the first set are based on the assumption that speech is the primary modality of an utterance, while gesture is subordinate and serves an auxiliary function; theories of the second set, in turn, consider speech and gesture to be equal partners in the production of multimodal utterances.

Chapter 3

Artificial Communicators: State of the Art

This chapter sets out to provide an overview of the state of the art of computational approaches aiming at the generation and evaluation of speech and gesture for artificial communicators. In this way, it establishes the technical context for the implementation presented in Part II as well as the empirical context for the experiments described in Part III of this thesis.

Two research areas are relevant to the present work: firstly, computer animation in which researchers have developed frameworks to realize multimodal communication behavior in virtual conversational agents; secondly, robotics in which researchers have explored various approaches to generate non-verbal behaviors along with speech in humanoid robots. The challenges are similar in that both research areas demand a high degree of control and flexibility so that human-like motion can be adapted to a system with non-human kinematics. The levels of complexity encountered in each field, however, are not equivalent. Although the range of different body types found in virtual embodied agents is manifold and hence challenging, character animation has less restrictive motion than even the most state-of-the-art humanoid robots (Pollard et al., 2002). For example, animation of virtual agents reduces or even eliminates the problems of handling joint and velocity limits; in a robot body, however, these have to be explicitly addressed given real physical restrictions.

The chapter is organized as follows. In Section 3.1 the first area of interest is explored by presenting and assessing various scientific frameworks that endow virtual agents with multimodal communicative behavior. In addition, evidence resulting from the empirical evaluation of such systems is provided. In Section 3.2 the second field of interest is introduced by presenting a number of related robotic systems from both technical and empirical points of view.

3.1 Virtual Agents

In contrast to the research area of robotics, the challenge of generating speech and co-verbal gesture has already been tackled rather extensively in various ways within the domain of virtual agents. Such agents, also known as *Virtual Humans* (*VHs*), *Intelligent Virtual Agents* (*IVAs*), and *Embodied Conversational Agents* (*ECAs*), have been around for over a decade. As a result, many technical implementations have been described in the relevant literature, some of which are presented in the following (3.1.1). Findings resulting from empirical studies that investigate the effect of such multimodal communicative behavior on human-computer interaction (HCI) using virtual agents are then summarized (3.1.2).

3.1.1 Technical Implementations

Existing implementations vary across the different levels of the behavior generation pipeline which has been introduced in the first chapter and is illustrated in Figure 1.3. Since the present thesis is concerned only with behavior realization at the lowest level of the pipeline, the description of existing frameworks in this chapter will mainly focus on differences at this level. Moreover, since the actual object of investigation is multimodal behavior realization for humanoid *robots*, the review of virtual agent platforms is not intended to be exhaustive. Rather, by outlining a selection of frameworks, it gives an idea of the beginnings of ECA research and the difficulties already encountered when generating synchronized speech and gesture for computer animated agents. This, in turn, will provide a general understanding of the increasing challenges that have to be tackled when realizing such behaviors on a physical robot platform.

Animated Conversation

The first ECA system to provide time-based synchronization of speech and gesture at the word and syllable level was introduced by Cassell et al. (1994). Drawing inspiration from psychological gesture research, their *Animated Conversation* system models the generation of both verbal and non-verbal behavior such as hand gesture, gaze, and facial expression in a hierarchical system architecture as illustrated in **Figure 3.1**. The multimodal capabilities of the system are demonstrated in an interaction between two autonomous graphical agents.

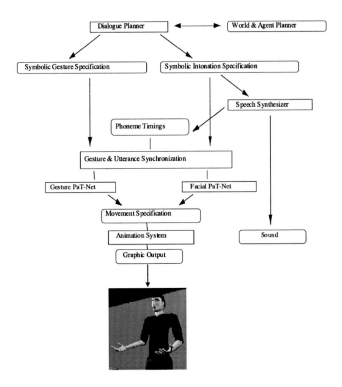

Figure 3.1: Architecture of the *Animated Conversation* system and generated example output (Cassell et al., 1994; Cassell, 1996).

At the top level of the system, the 'Dialogue Planner' assigns gestures to the rhematic content of speech, i.e., the part that provides some new information about the discourse theme. This process is based on a simple rule-based approach: if the semantic content of speech is literally or metaphorically spatial, an iconic or metaphoric gesture respectively is selected; if the content is spatializable or refers to an entity, a deictic gesture is selected; otherwise, if the new information in discourse falls into neither of these categories, a beat gesture is selected.

At the 'Gesture & Utterance Synchronization' level, the coordination of speech and gesture is performed heuristically by adapting gesture timing to phoneme time information derived from the speech synthesizer. That is, start and end times of phonemes are used to parametrize gesture time (Cassell, 1996). Apart from this synchronization step, speech synthesis and gesture generation are prepared in two separate processes that run independently from each other in a feedforward manner. Each gesture has a fixed duration and is generated from a library of predefined templates which is called upon by the animation system.

This approach has several limitations. First, the use of predefined gesture templates limits the flexibility and variety of generated co-verbal gesture. Additional gestures have to be hand-crafted and added to the lexicon together with the timing information for both the length of the preparation time required and the total execution time of the gesture. This fixed timing information leads to the second major limitation: given the incapacity to adjust gesture speed, the animation of a gesture can only be handled and scheduled as an atomic unit. If the total duration of a prospective gesture exceeds the time given by an intonation phrase, it is completely skipped. Finally, the separation and isolation respectively of the two planning processes for speech and gesture poses another issue: after scheduling a gesture for a spoken utterance, no feedback information about the current state of the animation system is sent back to the planning module, making subsequent adjustments impossible. This ballistic approach also affects the processing of subsequent multimodal utterances, which are prepared completely independently from each other. Ultimately, the Animated Conversation system suffers from a lack of real-time capabilities in the generation of multimodal behaviors and does not allow for interaction with a real user (Noma et al., 2000; Cassell et al., 2000).

REA

A further approach introduced by Cassell et al. (2000) is the *REA* (Real Estate Agent) system in which a female virtual agent operates as a real estate salesperson. While interacting with a human user, the agent presents various real estate properties and provides detailed visual and verbal information about a particular property upon the user's request. The system builds on experience gained from the Animated Conversation project, however, it is more advanced in that it can handle bi-directional communication. That is, in addition to the real-time generation of multimodal utterances, the system also attempts to process and understand multimodal input from the user.

The underlying system architecture (see **Figure 3.2**) is based on sequential processing of user input and provides different modules for each processing stage of

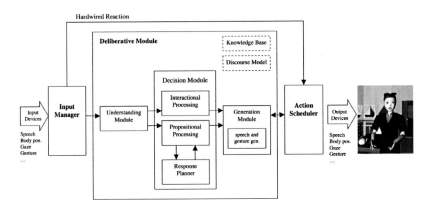

Figure 3.2: Architecture of the *REA* system and generated example output (Cassell et al., 2000).

the discourse. First, the 'Input Manager' module handles input data from various modalities, as collected by two video cameras mounted above the agent's screen and by a microphone attached to the user. The module then decides whether the data requires an instant, hardwired reaction, e.g., greeting behavior to stimuli such as the appearance of the user, or deliberative discourse processing.

The 'Deliberative Module' handles input from both interactional behaviors and propositional behaviors. Given such input, it maintains a discourse model for the interpretation of the current conversation, e.g., by keeping records of previous statements made by both the user and the agent. For this purpose, it employs an 'Understanding Module' and a 'Decision Module' with reference to a global knowledge base. This, in turn, results in a formulated sequence of actions that express the desired communicative or task goal. The 'Generation Module' subsequently translates the chosen discourse functions into surface behaviors by producing a set of primitive actions including speech, gesture, and facial expression.

Finally, and most importantly with regard to the present work, the 'Action Scheduler' module represents the motor controller in which the multimodal output actions for the embodied agent are coordinated and scheduled at the lowest level. During the process, the module falls back on a set of atomic modality-specific commands and subsequently executes them in a synchronized manner. To achieve such cross-modal synchrony, gesture timing is heuristically adapted to the timing of ballistically generated speech, in a similar fashion as realized in the Animated Conversation system. Furthermore, execution of behaviors is event-driven, i.e., event conditions are assigned to each output action, specifying when exactly and in what interdependence the action should be executed. Besides the scheduler, the output system further consists of 'Output Devices' comprising an animation component as well as a rendering component.

Despite its elaborate architecture, the REA system is still limited in some aspects. Cassell et al. (2000) particularly point out major weaknesses with regard to timing as well as to synchrony of multimodal behaviors. Firstly, the short response time needed for hardwired reactions stands in stark contrast to the great latency caused by the deliberative discourse processing module, leading to inconsistencies in the agent's overall behavior. Secondly, synchronization of verbal and non-verbal behaviors is not always successfully accomplished by the system, e.g., a hand gesture may occur *after* its co-expressive speech segment has been uttered. As stated by the authors, "the problem is due primarily to the difficulty of synchronizing events across output devices, and of predicting in advance how long it will take to execute particular behaviors" (Cassell et al., 2000, p. 59). Adding to this issue is the fact that gestures are handled as atomic predefined keyframe animations which, apart from some open parameters, cannot be flexibly timed and adjusted, e.g., to varying speed demands. Finally, once triggered by a preconditioned event, both verbal and non-verbal behaviors are executed ballistically, i.e., the system does not provide any means for feedback-based adjustments of miscalculated scheduling.

BEAT

Yet another system proposed by Cassell et al. (2001) is the Behavior Expression Animation Toolkit (*BEAT*). Based on a modular architecture (see **Figure 3.3**), the system enables an animated virtual human to generate appropriate and synchronized non-verbal behaviors and synthesized speech from typed input text. For this purpose, it draws upon rules derived from research in human conversational behavior which are represented in a knowledge base as well as in a discourse model. More specifically, the processing of the user's input is performed in real-time by three major modules which are further illustrated in the following.

First, with reference to a discourse model, the 'Language Tagging' module

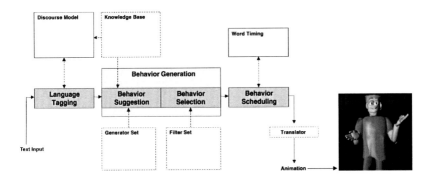

Figure 3.3: Architecture of the *BEAT* system and generated example output (Cassell et al., 2001).

analyzes the plain-text input and annotates it with XML tags indicating linguistic and contextual features. For this, the text is first broken up into clauses, each of which is considered as representing a proposition. Clauses are typically recognized by the placement of verb phrases or punctuation. Based on heuristics proposed by Hiyakumoto et al. (1997), each clause is divided into smaller units of information structure to distinguish the discourse theme (i.e., the present topic) from the rheme (i.e., new subject or information). These units, in turn, are further broken down into word phrases which either describe an action or an object. Finally, at the word level, new words that have not previously occurred in the discourse as well as contrasting words are tagged according to these additional properties.

The resulting XML trees are subsequently forwarded to the 'Behavior Generation' module which consists of two sub-modules. First, the 'Behavior Suggestion' module which uses a set of rule-based behavior generators augments the XML text with suggestions for appropriate non-verbal behaviors. Each of these generators is responsible for a different kind of behavior, e.g., the 'beat gesture generator' suggests beat gestures for rhematic discourse information when no other gesture is found to be appropriate. Resulting behavior suggestions are specified with a tree node that defines the time interval in which they are active, a priority value that can be used to resolve potential conflicts, degrees of freedom (DOF) required for animation, and the gesture specification needed for rendering. The module also determines whether different behavior suggestions can co-articulate, i.e., occur during other behaviors using the same DOF. Given the updated tree which contains many potentially incompatible behavior suggestions, the 'Behavior Selection' module then applies a set of filters to cut down the choice of behaviors to the set of behaviors that eventually will be used for animation. For this, filters can delete conflicting suggestions or those with a priority falling below a predefined threshold.

Finally, in the last step of processing, the 'Behavior Scheduling' module converts the incoming XML tree into a set of time-based instructions to be executed by the animation system. Cross-modal synchrony is achieved by constructing a schedule for the animation subsystem based on word and phoneme timing information obtained from the text-to-speech (TTS) engine. At the final stage of scheduling, the abstract animation schedule is translated into a format specific to the animation subsystem in use, typically defined as a continuous keyframe animation.

Although the toolkit is extensible, e.g., by allowing for new generator rules as well as predefined gestures to be later added to the system, it has similar limitations as its predecessors. Again, the level of animation is restricted to lexicon-based, predefined animations of set durations. This constrains the flexibility of the system given its inability to adapt the gesture's execution time, e.g., by stretching or shortening it. At the same time, synchronization is a unidirectional process in which synthesized speech cannot be modulated to adapt to gesture timing. Finally, as in the previously presented systems, speech and gesture are executed ballistically in BEAT, which prevents further adjustments once behaviors have been scheduled and sent to the realization pipeline.

GRETA

The interactive expressive *GRETA* system (Hartmann et al., 2002, 2005) is another example of a real-time animated conversational agent. The female character is able to communicate with a human user by means of verbal and non-verbal behaviors such as gestures, head movements, gaze, and facial expressions. In view of the present work, the system's modular *Gesture Engine* is of particular interest and is illustrated in **Figure 3.4**. The engine interprets an utterance file specified in the Affective Presentation Markup Language (*APML*; DeCarolis et al., 2004) and generates the designated multimodal behavior on the GRETA agent.

For this, specified speech output is first processed by the TTS engine *Festival* (Black and Taylor, 1997) to obtain phoneme timing information which is required to

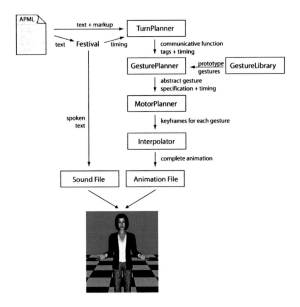

Figure 3.4: Architecture of the *Gesture Engine* of the *GRETA* system and generated example output (Hartmann et al., 2005).

synchronize gesture to speech. Markup tags describing communicative functions are considered for gesture matching and are extracted by the 'TurnPlanner'. Based on a straightforward lexical lookup, the 'GesturePlanner' then assigns predefined prototype gestures stored in the 'GestureLibrary' to the communicative function tags. Given their abstract specification and timing information, the 'MotorPlanner' concretizes the gesture prototypes by calculating joint angles and timing for each keyframe of the gesture animation. Finally, a set of 'Interpolators' generates intermediate frames for the complete animation file which is then executed concertedly with the synthesized sound file.

Similar to the above described virtual agent systems, the GRETA framework is limited in that gesture timing is solely determined by speech timing. To account for this limitation, gesture preparation and retraction phases are to some extent adjustable within their predefined time ranges, and intermediate rest poses are inserted if the time between two gestures is too long for a direct transition. In addition, the prototype gestures stored in the gesture library can be parametrized at different levels of the planning pipeline, e.g., by modulating the spatial or temporal extent, or the smoothness and continuity of the movement (Hartmann et al., 2005). Nevertheless, the underlying scheduling algorithm is completely deterministic. This means, all time constraints are set a priori, i.e., before the actual behavior execution which is then performed ballistically with no possibility of on-line adjustment. This results in limited flexibility of the system, which is further fortified by the lexicon-based storage and usage of predefined gesture templates.

MAX

In contrast to all of the above described approaches, the system underlying the virtual agent *MAX* (Multimodal Assembly eXpert) is the first to provide for mutual adaptation mechanisms between the timing of speech and gesture (Kopp and Wachsmuth, 2004). It further builds upon an integrated architecture in which the planning of both content and form across both modalities is coupled (Kopp et al., 2008), taking into account the meaning conveyed in non-verbal utterances.

Operating as the agent's real-time behavior realizer, the *Articulated Communicator Engine* (ACE) aims at generating lifelike, synchronized verbal and non-verbal behaviors in a natural flow of multimodal behavior. ACE has been employed in a number of research projects and is currently used as an action generation framework in multiple virtual agent systems besides MAX, for example, NUMACK (Kopp et al., 2004).

The multimodal production model of ACE draws inspiration from McNeill's *Growth Point Theory*[1] of human communication (Sowa et al., 2008), according to which speech and gesture are tightly coupled and co-express the same *idea unit*[2] (McNeill, 1992). It is further based on an empirically suggested assumption referred to as the *segmentation hypothesis* (Kopp and Wachsmuth, 2004), which claims that the production of continuous speech and gesture is organized in successive segments. Each of these segments, in turn, corresponds to what has been previously defined as a multimodal *chunk* of speech-gesture production (see Definition 2, Chapter 2.2.1), comprising a pair of an intonation phrase and

---

[1]See Chapter 2.2.3
[2]See Chapter 2.2.1 and Figure 2.4

a co-expressive gesture phrase. That means, complex utterances with several different gestures consist of multiple successive chunks.

As a result of this approach, the innovative quality of the ACE framework particularly lies in its incremental on-line scheduling of multimodal behavior for virtual agents, which handles cross-modal interactions at different levels of an utterance. This corresponds to mutual adaptations that are believed to take effect when humans synchronize co-expressive elements of the two modalities (Sowa et al., 2008). The modular ACE system architecture implementing this theory-driven approach is illustrated in **Figure 3.5** and is briefly elucidated in the following. More detailed information about the underlying mechanisms and relevant implementation details are further provided in Chapter 4.2.1.

It is assumed that, before ACE comes into play at the behavior realization level, the selection of appropriate actions has been taken care of by other, higher-level,

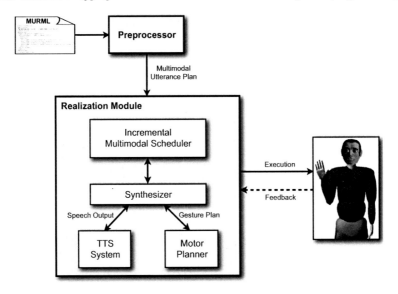

Figure 3.5: Architecture of the *Articulated Communicator Engine* (ACE) of the *MAX* system and generated example output.

planners of the behavior generation pipeline[3]. The resulting multimodal behavior description is converted into a format that can be processed by ACE. To this end, a specification file is generated using the XML-conform Multimodal Utterance Representation Markup Language (*MURML*; Kranstedt et al., 2002), based on which gestures can be specified in two different ways.

Firstly, similar to the first three virtual agent systems described in this section, gestures can be described in terms of keyframe animations. For this, each keyframe specifies a part of the overall gesture movement pattern by describing the state of each joint at a given time. Speed information for the interpolation between every two successive keyframes and the corresponding affiliation to parts of speech are obtained from assigned time identifiers. For use in ACE, keyframe animations can be defined either manually or derived from motion capturing data from a human demonstrator, allowing for the animation of virtual agents in real-time.

Secondly, gestures can be specified on a higher level of abstraction in a feature-based representation format adopted from HamNoSys (Prillwitz et al., 1989), a notation system for sign languages. This approach is based on the assumption that the communicative intent of a gesture can be sufficiently described in terms of spatio-temporal form features, corresponding to the posture of the meaning-bearing stroke phase (Kranstedt et al., 2002). As with keyframe animations, gesture affiliation to dedicated linguistic elements is determined by matching time identifiers. **Figure 3.6** illustrates an example of a feature-based MURML specification for multimodal behavior generation, describing the co-verbal gesture in an overt form. In the present work, the focus is on the generation of such feature-based utterance descriptions, since this high-level definition of gestures in terms of their spatial targets and form properties appears to be more biologically plausible than the keyframe-based approach (see Chapter 4.2.1). Moreover, from a semantic point of view, specifying gestures based on their overt form features reflects a more conceptual approach to describing meaningful gestural body movements.

Once the ACE system has received the MURML input file, it is translated into a multimodal utterance plan by the 'Preprocessor'. The utterance plan is then

---

[3]See Kopp et al. (2008) and Bergmann and Kopp (2009) for more information on how this is currently conducted in the MAX system. Note that it is also possible to refer to predefined gestures stored in a 'gesticon', however, this is not a favored strategy, since it limits the flexibility of the agent's behavior.

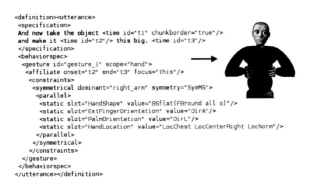

```
<definition><utterance>
 <specification>
 And now take the object <time id="t1" chunkborder="true"/>
 and make it <time id="t2"/> this big. <time id="t3"/>
 </specification>
 <behaviorspec>
  <gesture id="gesture_1" scope="hand">
   <affiliate onset="t2" end="t3" focus="this"/>
    <constraints>
     <symmetrical dominant="right_arm" symmetry="SymMS">
      <parallel>
       <static slot="HandShape" value="BSflat(FBround all o)"/>
       <static slot="ExtFingerOrientation" value="DirA"/>
       <static slot="PalmOrientation" value="DirL"/>
       <static slot="HandLocation" value="LocChest LocCenterRight LocNorm"/>
      </parallel>
     </symmetrical>
    </constraints>
  </gesture>
 </behaviorspec>
</utterance></definition>
```

Figure 3.6: Example of a feature-based MURML specification for multimodal utterances and the resulting gesture.

forwarded to the actual 'Realization Module' which contains the 'Incremental Multimodal Scheduler & Synthesizer'. This sub-module, in turn, which has access to the text-to-speech system and the motor planner, plans and schedules the behaviors of the two modalities so that they can be appropriately synchronized during execution. The incremental scheduling and production of successive coherent chunks is realized by processing each chunk on a separate blackboard running through a sequence of different states as illustrated in **Figure 3.7**. Given an utterance consisting of multiple chunks, the modalities are coordinated at two different levels.

First, for intra-chunk scheduling, the gesture is timed such that its meaning-bearing stroke phase starts before (0.3 seconds or one syllable) or with the linguistic affiliate and completely spans it, if necessary by inserting a dedicated hold phase. This temporal synchrony is mainly accomplished by the gesture's adaptation to the timing of speech, for which absolute phoneme timing information is obtained from the text-to-speech system. Once scheduled and ready to be executed, speech and gesture run ballistically within a chunk, i.e., their execution is unaffected by the progress of the other respective modality.

Second, for inter-chunk scheduling between two successive chunks, both speech and gesture can anticipate the forthcoming chunk and adapt to it depending on their timing. For gesture, this adaptation referred to as *co-articulation effects* may range from the insertion of an intermediate rest pose to a direct transition

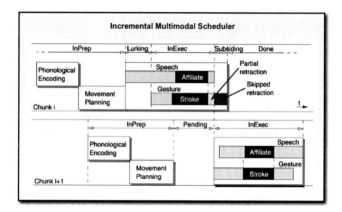

Figure 3.7: Incremental multimodal scheduler in ACE running through a sequence of processing states (adapted from Kopp and Wachsmuth, 2004).

movement skipping the retraction phase or parts thereof. For speech, a silent pause may be inserted between two intonation phrases for the duration of the preparation phase of the following gesture. Such adaptive flexibility in the form of inter-chunk synchrony is achieved by a global scheduler which plans the following chunk in advance while monitoring the chunk currently in execution. Once the predecessor has been executed, the plan for the next chunk may be refined based on the current state of speech and gesture generation. It is not until this moment that animations satisfying movement and timing constraints now determined are created. This highly flexible level of generativity builds upon the ability of ACE to generate all animations that are required to drive the agent's skeleton in real-time and from scratch (Kopp and Wachsmuth, 2004).

In summary, the ACE system attempts to overcome some of the issues found in the previous frameworks presented above, based on the implementation of several innovative features. Specifically, these are expressed in the level of flexibility provided by the ACE scheduler as well as the possibility to generate gestures without the use of predefined templates or detailed keyframe descriptions. However, despite the increased degree of flexibility compared to the above described frameworks, even in ACE the generation of speech and gesture suffers from some conceptual shortcomings. In particular, the cross-modal synchronization mechanisms realized in the framework are not entirely realistically modeled. Despite providing for mutual adaptation at the inter-chunk level, lacking adjustability within a chunk as well as the ballistic generation of complete gesture and intonation phrases conflict with findings from psychology (e.g., Kendon, 2004).

Addressing the challenge of flexibly scheduling and synthesizing multimodal behavior, a few approaches using the newly established *Behavior Markup Language* (BML; Kopp et al., 2006; Vilhjálmsson et al., 2007) have been introduced in recent years, for example, *SmartBody* (Thiebaux et al., 2008), *EMBR* (Heloir and Kipp, 2010), and *Elckerlyc* (van Welbergen et al., 2010). Nevertheless, even at the present time, ACE is still considered a very sophisticated framework for the animation of virtual agents and is frequently extended with new features, including a fusion with the Elckerlyc system for support of the BML standard. As a result, ACE represents a suitable foundation for a speech-and-gesture generation system for humanoid robots as envisaged by the present work.

3.1.2 Empirical Evaluations

In addition to the technical contributions presented in the area of embodied conversational agents, there has also been active work in evaluating the effect of multimodal communicative behavior in virtual agents. Several empirical studies have focused on non-verbal behaviors such as gaze, nodding, or facial expressions, partly in combination with emotions (e.g., Heylen et al., 2002; Buisine et al., 2006; Kipp and Gebhard, 2008; Lee et al., 2010). Of more relevance to the present work, however, are those studies that have investigated the effect of hand and arm gestures in particular on human perception of virtual agents, e.g., with regard to traits such as naturalness, believability, or likability.

In one such study, Cassell and Thorisson (1999) showed that a virtual agent's non-verbal behaviors, specifically beat gestures, are more important to users in conversational interaction than the expression of emotional feedback. Besides playing a crucial role in supporting the dialog process, such non-verbal behaviors were also found to result in higher user ratings of the agent's perceived lifelikeness and fluidity of interaction.

In contrast, Krämer et al. (2003) found no significant effect on humans' perception of the agent in a comparison between an ECA using gesture and one that did not gesture while presenting a TV/VCR device to the user. The gesturing agent was perceived as equally likable, competent, and relaxed as its non-gesturing counterpart. However, a significant effect was found when comparing participants who were confronted with a system that used only written text output or synthesized speech output with those confronted with an embodied virtual agent: while participants in the text and audio condition reported increased helpfulness of the non-animated system, participants confronted with the virtual agent felt more entertained.

In a study using various animated agents for short technical presentations (Buisine et al., 2004), three different speech-gesture cooperation strategies with regard to the semantic relationship between both modalities were tested: redundancy, complementarity, and speech-unrelated gestures. Multimodal strategies proved to influence participants' subjective impressions of the quality of explanation, which was particularly true for male participants: the quality of the agent's explanation was rated more positively when it was using redundant or complementary gestures as opposed to speech-unrelated gestures. In addition, the agent's appearance was found to affect its likability, as well as participants' ability to recall the content presented to them by the agent.

Kipp et al. (2007) applied the concept of gesture units (Kendon, 1980) to a virtual agent's gesture generation, in order to produce a continuous flow of movement. This was realized by connecting neighboring gestures within a single utterance segment with a hold rather than a retraction phase. They hypothesized that a virtual agent generating such gesture units is perceived as more natural than one using singletons by separating every two gestures with a retraction. Results confirmed their hypothesis, as participants not only rated the agent producing gesture unit as more natural, but also as more friendly, more trustworthy, and, by trend, more competent. Finally, the agent using singleton gestures was perceived as more nervous than the one employing hold phases between successive gestures.

In another study conducted by Krämer et al. (2007), the conversational agent MAX communicated by either utilizing a set of co-verbal gestures alongside speech, typically by self-touching or movement of the eyebrows, or by utilizing speech alone without any such accompanying gestures. Human participants were then invited to rate their perception of MAX's behavioral-emotional state, for example, its level of aggressiveness, its degree of liveliness, etc. Crucially, the results of the study suggested that virtual agents are perceived in a more positive light when they are able to produce co-verbal gestures alongside speech rather than acting in a speech-only modality.

Bergmann et al. (2010) modeled the gestures of MAX based on real humans' non-verbal behavior and subsequently set out to question the communicative quality of these models via human participation. The main finding was that MAX was perceived as more likable, competent, and human-like when gesture models based on individual speakers were applied, as opposed to combined gestures of a collection of speakers, random gestures, or no gestures.

Finally, Neff et al. (2010) investigated how language generation, gesture rate, and different movement performance parameters can be varied to increase or decrease the perceived extraversion of a virtual agent. Testing four experimental conditions, these factors were modulated to change the agent's level of extraversion (from very low to very high), demonstrating indeed a statistically significant effect on human perception of this personality trait.

3.2 Robotics

While being a fairly established subject of research in the field of conversational virtual agents, the generation together with the evaluation of the effects of co-verbal communicative gesture is still largely unexplored in robotics. Typically, in traditional robotics, recognition rather than synthesis of gesture has been mainly brought into focus. Furthermore, within the few existing approaches claiming to be dedicated to gesture generation, the term 'gesture' has been widely used to denote object manipulation tasks rather than non-verbal communicative behaviors. For example, Calinon and Billard (2007) refer to the drawing of stylized alphabet letters as gestures in their work.

Many researchers have focused on the translation of human motion for gesture generation in various robots, usually aiming at imitation of movements captured from a human demonstrator, e.g., Billard et al. (2008). Further techniques for limiting human motion of upper body gestures to movements achievable by a variety of different robotic platforms have been presented by Pollard et al. (2002) and Miyashita et al. (2006). However, these models mainly focus on technical aspects of generating robotic motion that fulfills little or no communicative function. In addition, they are limited in that they do not combine generated non-verbal behaviors with further output modalities such as speech.

Generally, only few approaches have so far pursued both the generation of humanoid robot gesture and the investigation of human perception of such robot behavior; however, an upward trend has been noticeable in recent years. In the following, several technical approaches implementing speech-gesture generation for robots, as found in the relevant literature, are presented (3.2.1). Findings from empirical studies that investigate the effect of such multimodal communicative behavior on human-robot interaction (HRI) are then summarized (3.2.2).

3.2.1 Technical Implementations

Only a few robotic systems incorporate both speech and gesture synthesis; however, in most cases the robots are equipped with a set of predefined or even pre-recorded gestures that are not generated on-line but simply replayed during human-robot interaction. Furthermore, such approaches are often realized on rather simple robotic platforms with less complex robot bodies, e.g., providing only limited mobility and less DOF. Characteristic features of related approaches to robot gesture generation are categorized in the following and are further illustrated by presenting a selection of implementation examples that fall into each category. Note, however, that most outlined systems belong to multiple of the listed categories.

Predefined gestures and limited body expressiveness

The personal robot *Maggie* (see **Figure 3.8a**) aims at interacting with humans in a natural way in order to establish a peer-to-peer relationship with them (Salichs et al., 2006; Gorostiza et al., 2006). For this purpose, the robot is equipped with a predefined set of implemented gestures, which can be expressed through movements of the head, the eyelids, and the arms. However, given the design of the robot's arms with only 1DOF each and no end-effectors (i.e., hands) attached to them, the level of arm gesture expressiveness is rather limited. For example, the generation of iconics and other complex communicative gestures typically requires more DOF than provided by the Maggie platform.

Another example of gesture generation presented by Sidner et al. (2003) is the penguin robot *Mel* (see **Figure 3.8b**), which is able to engage humans in a collaborative conversation. In a dedicated research scenario, Mel employs speech and gesture to indicate engagement behaviors while telling human interlocutors about the functions of a newly invented object placed on a table in the same room. However, the gestures generated in this context are also predefined in a set

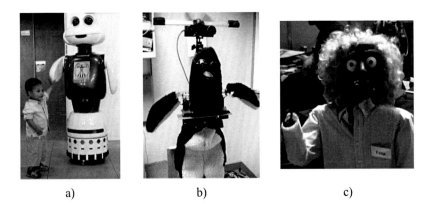

a)   b)   c)

Figure 3.8: Examples of robotic platforms used for gesture generation: a) personal robot *Maggie* (Salichs et al., 2006), b) penguin robot *Mel* (Sidner et al., 2003), c) communication robot *Fritz* (Bennewitz et al., 2007).

of action descriptions, called the 'recipe library', e.g., specifying how to greet a visitor, how to present the invented object, etc. The descriptions are not written in scripts, but rather as task models with annotations of how to convey certain utterances. Besides lacking on-line generation capabilities, the zoomorphic design of the Mel platform provides only limited flexibility and expressiveness with regard to feasible body movements for arm gesture generation.

The communication robot *Fritz* (see **Figure 3.8c**) introduced by Bennewitz et al. (2007) uses speech, facial expressions, eye-gaze, and gesture to appear livelier while interacting with people. Although the arm gestures produced during interactional conversations are generated on-line, they mainly consist of predefined culture-specific emblems, beat and pointing gestures to direct the attention of the communication partner towards certain objects. To further add to the robot's liveliness, Fritz performs minuscule non-gestural arm movements in randomized oscillations instead of standing still while idle. However, body movements are performed in a rather jerky fashion. Furthermore, and similar to the previously described platforms, the level of gesture expressiveness is limited since the robot is not equipped with any hands and generally comprises less DOF.

Unfavorable appearance of the gesturing robot

The appearance of the gesture generating robots presented in the previous subsection differs substantially from the humanoid robot used for the present work. Besides comprising less DOF and lacking end-effectors, they generally expose only little or no humanoid traits. As stated by Minato et al. (2004), however, the appearance of a robot can be just as important as its behavior when evaluating the experience felt by human interaction partners. In other words, the robot's design is crucial if we are to eventually study the effect of robot gesture on humans.

Macdorman and Ishiguro (2006) have researched human perception of robot appearance as based on different levels of embodiment, with android robots representing the most anthropomorphic form. Although an innovative approach, android robots only feature certain hard-coded gestures and thus still lack any real-time gesture-generating capability. Moreover, findings presented by Saygin et al. (2011) suggest that the mismatch between the highly human-like appearance of androids and their mechanical, less human-like movement behavior may lead to increased prediction error in the brain, possibly accounting for the 'uncanny valley' phenomenon (Mori, 1970).

Given these empirical findings, a major advantage of using the Honda humanoid robot as a research platform for the present works lies in its humanoid, yet not too human-like appearance and smooth, yet not completely natural movement behavior. Although the robot in question cannot mimic any facial expression, it is advantageous to use such a platform, since the focus of the present work lies on hand and arm gestures. This way, the perception of the robot's gestural arm movements can be assessed as the primary non-verbal behavior.

Limitation to a single gesture type

Many approaches of communicative gesture synthesis for humanoid robots are limited to the implementation and evaluation of a single type of gesture, instead of providing a general framework that can handle all, or at least a variety of, gesture types. For example, Sugiyama et al. (2007) and Okuno et al. (2009) present systems that focus on the generation of deictic gestures for object indication and route direction-giving respectively, both using the *Robovie* robot (see **Figure 3.9a**). Okuno et al. justify their choice to focus only on one type of gesture, namely deictics, by arguing that it is difficult for a robot to express other gesture types

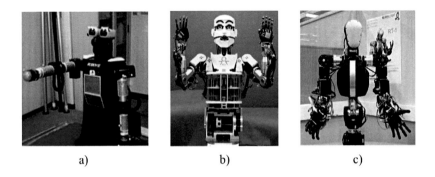

a)   b)   c)

Figure 3.9: Examples of robotic platforms used for single gesture type generation: a) deictic gesture performed by *Robovie* (Okuno et al., 2009), b) emblematic gesture for expressing the emotional states of the *WE-4RII* robot (Itoh et al., 2004), c) open hand gesture generated by *BERTI* (Bremner et al., 2009).

such as iconics or beats. They further explain that the effect of such imagistic or rhythmic gestures on the human listener are unclear and thus declare deictics to be "the most promising gesture" (p. 56).

In another approach targeting only one gesture type, Itoh et al. (2004) implemented emblems for displaying emotional states on the *WE-4RII* robot (see **Figure 3.9b**). For this, gesture generation is based on predefined motion patterns specifying the position and posture of the robot's hands, as well as time information about the trajectory using 3D spline functions. However, since emblematic gestures typically communicate without the need for spoken language, speech synthesis and synchronization with the generated gestures were not addressed.

Finally, Bremner et al. (2009) focused on a small range of gestures of the open hand for the upper-torso humanoid robot *BERTI* (see **Figure 3.9c**). To verify and evaluate their proposed control algorithm, they implemented four different open-handed gestures. Timings for speech-gesture synchronization were guessed and predefined at the coding stage, thus resulting in gesture movements that are entirely pre-planned in form and duration. During execution, movements of the different gesture phases (i.e., preparation, stroke, and retraction) are triggered by events in the speech. Although open hand gestures may be employed to generate a variety of emblems and beat gestures, the exclusive use of only one hand shape significantly limits the flexibility and expressiveness of the gesturing robot.

Building on experience from the virtual agents domain

In recent years, the idea of building upon experiences gained from previously developed virtual agent frameworks has also been taken up by a few other researchers. For example, Mead et al. (2010) generate gestural behaviors along with speech for the upper-torso humanoid robot *Bandit III* (see **Figure 3.10a**) using the NonVerbal Behavior Generator (NVBG), a system that was originally developed for embodied conversational agents (Lee and Marsella, 2006). NVBG consists of a set of rules that can be employed to determine which gestures correspond to a given verbal content. These rules are derived from user activity modeling and are based on specific keywords and sentence structures. In contrast to the work presented in the present thesis, Mead et al.'s approach uses an element from the virtual agent domain for behavior *planning* rather than *realization* at the lowest level of the behavior generation pipeline (see Figure 1.3). In their proposed system, the actual realization of selected gestural behaviors is performed by a robot-specific implementation that roughly attempts to synchronize robot gestures to the speech output at the phrase level rather than to the exact affiliate.

More similar to the present work in that it transfers a virtual agent framework for behavior realization to a humanoid robot is the approach proposed by Le et al. (2011). By building on and extending the above described GRETA system (see Figure 3.4), the resulting speech-gesture generation system aims at endowing the robot *NAO* (see **Figure 3.10b**) with multimodal communicative expressiveness.

a)

b)

Figure 3.10: Examples of robotic platforms used for gesture generation: a) *Bandit III* robot (Mead et al., 2010), b) *NAO* robot (Le et al., 2011; Gouaillier et al., 2009).

For this, the original framework was extended with a gesture repository containing pre-tested gesture descriptions specifically compiled for the NAO robot, i.e., taking into account feasibility constraints of the robot's physical body. Gesture specifications are stored in a symbolic format, while the generation of the actual gestures and their trajectories respectively is initiated and realized at run-time. However, the limitations of the underlying virtual agent framework as highlighted in Section 3.1.1, e.g., the need for predefined gestures stored in a lexicon as well as the lack of cross-modal synchronization mechanisms, still remain in the proposed robot-specific system. In addition, with a height of only 0.57m (Gouaillier et al., 2009), the NAO robot used in Le et al's approach is fairly small and thus only limitedly suitable for operating and interacting in the human living space.

Open-loop control and unidirectional synchronization

One of the few models that resembles our approach in that it attempts to generate a multitude of gesture types for the Honda humanoid robot was presented by Ng-Thow-Hing et al. (2010). Similar to the above described BEAT system (Section 3.1.1), their proposed model reconstructs the communicative intent through text and parts-of-speech analysis to select appropriate gestures based on arbitrary input text. For the generation of a selected gesture, predefined gesture templates are used, specifying the basic trajectory in a set of key points for the hand position, wrist rotation, and hand shape. These parameter values can be further modulated based on probabilistic elements, so that multiple instances of the same input text do not result in the generation of identical gestures.

However, a major limitation of this system – which also applies to all other approaches presented in this section – is the open-loop control of the implementation. That means, once the gesture plan has been generated from a given input text, speech and gesture are generated ballistically and cannot be further re-adjusted at run-time, e.g., based on sensory feedback. Furthermore, as it is generally the case in systems that attempt to synchronize synthetic speech and gesture, running speech completely dictates the timing of the generated gestures, while gestures, in contrast, cannot affect the produced speech output.

To the best of knowledge, such unidirectional synchronization as well as the open-loop production of multimodal behaviors is characteristic of all currently existing approaches to the generation of robot gestures. As a result, the major technical contribution of the work presented in this thesis is twofold: first, it provides an implementation of a speech-gesture production system for a humanoid robot which is based on one of the most sophisticated virtual agent frameworks; second, it stands out from other existing robotic gesture generating systems by incorporating reactive closed-loop feedback for a more flexible approach to multimodal robot expressiveness (see RQ2).

3.2.2 Empirical Evaluations

Despite the interesting implications of the evaluation studies conducted with virtual agents, one must be cautious when transferring empirical findings from the domain of such animated graphical characters to the domain of social robots. Firstly, the presence of real physical constraints in robotic platforms can alter the perceived level of realism. Secondly, given the greater degree of embodiment that is possible in a real-world system, interaction with a robot is potentially richer: human participants could, for example, walk around or even touch a real robot. This makes the interaction experience more complex and is naturally expected to affect the outcome of the results.

Since the research area of speech-gesture synthesis for communicative robots is a fairly new one, most existing approaches are still mainly concerned with the technical realization of such systems. In the area of human-robot interaction, comparatively much research, e.g., carried out by Mutlu et al. (2009), has studied the effect of robot gaze as an important aspect of non-verbal behavior. In contrast, not much research has focused on the evaluation of co-verbal hand and arm gestures in particular. Consequently, only few data derived from empirical analyses of the effects and acceptance of communicative robot gesture in HRI have been provided in the literature so far. In many cases only pilot studies with few participants and no experimental hypothesis have been conducted, mainly to validate the technical implementation. In other instances, the evaluation of the system was merely undertaken by means of video-based studies in which participants did not actually interact with the robot, as done for example by Narahara and Maeno (2007), Li et al. (2009), Riek et al. (2010), and Ng-Thow-Hing et al. (2010).

However, in order to obtain a representative assessment of robot gesture and the human perception thereof, it is important to evaluate such non-verbal behavior in actual interaction scenarios. For example, a human-sized robot performing jerky gesture movements may potentially be perceived as dangerous by a human sharing the same interaction space, whereas a video of the same gesturing robot may not elicit any feelings of threat or fear in the human observer. Therefore, scope and space of robot gestures can only be accurately and appropriately observed and assessed based on real interactions in suitable test scenarios.

Up to the present time, only very few of such experimental HRI studies focusing specifically on the effect and perception of robotic hand and arm gestures have been conducted. In one such study, Sugiyama et al. (2007) investigated the degree to which participants perceived the interaction with the gesturing Robovie robot (see Figure 3.9a) to be as natural as inter-human communication. During the experiment, participants were asked to arrange five objects of similar size and shape freely within the interaction space and subsequently to point out one of them to the robot using a pointing gesture and a verbal cue. The robot then confirmed that it had recognized the according object by also generating a pointing gesture and a verbal cue. After a total of four trials, a questionnaire item ('yes/no' question) was used to ask participants whether the interaction with the robot was natural, which resulted in a significant majority of affirmative answers. However, although providing some useful findings, drawbacks of the study can be seen in the fact that it lacks a control condition and in the exclusive use of pointing gestures.

Another empirical study, conducted by Kim et al. (2008), examined how certain characteristics of the displayed robot gestures affect participants' impressions of the robot's personality. Specifically, gesture size (small vs. large), velocity (slow vs. fast), and frequency (low vs. high) were defined and manipulated as independent variables, yielding a combinatory sample of eight different gestures. In the experiment, participants were presented with these gestures performed by the *AMIET* robot (see **Figure 3.11a**) along with generated speech and, after each multimodal utterance, were asked to assess the robot's personality. This was conducted by rating two questionnaire items on seven-point Likert scales, one with endpoints 1 = introvert and 7 = extravert, the other one with endpoints 1 = feeling and 7 = thinking. To summarize the results, manipulation of gesture size, velocity, and frequency were shown to significantly influence the perceived level of introversion/extraversion, while a significant interaction effect of gesture size and velocity was found with regard to the personality item feeling/thinking. To illustrate, large and fast gestures of high frequency, for example, were associated with a more extroverted personality than smaller-sized, slow gestures of low frequency. In conclusion, the results indicate that controlling certain design factors of robot gestures affects human perception of a robot's

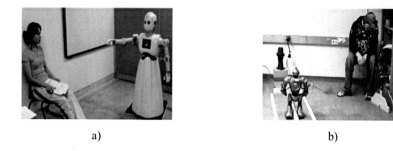

a)   b)

Figure 3.11: Robotic platforms used for evaluation studies on gesture-based HRI: a) *AMIET* robot (Kim et al., 2008), b) *Robosapiens RSMedia* (O'Brien et al., 2011).

personality traits and, at a more general level, suggest that a robot's personality can be successfully expressed via gestures. Although testing a variety of gesture types, a main limitation of the study, however, lies in the fact that participants did not actually interact with the robot but merely acted as passive observers during the experiment.

O'Brien et al. (2011) conducted an experimental study using the humanoid toy robot *Robosapiens RSMedia* (see **Figure 3.11b**), in order to evaluate the impact of the robot's gaze and arm gestures on a collaborative HRI task. For this, two conditions were designed: first, a gesture condition in which the robot used gaze (by orienting the head toward the human) and pointing gestures to refer to objects along with speech; second, a control condition in which neither the robot's head nor arms were moving, but rather remained in their default position for the entire interaction. Participants in both test groups were asked to solve a task together with the robot which, for example, asked the human to retrieve, use, or manipulate various objects in the interaction space. Analyses of post-experimental questionnaires revealed that the robot was rated as having significantly better interaction skills and as being a better collaborator when it interacted multimodally via speech, gaze, and gestures than when it used speech only. Although the study contributes some important findings, the individual effects of gaze and gesture respectively remain unidentified. Moreover, similar to the study by Sugiyama et al. (2007), only one type of gesture, namely deictics, was examined. Finally, given its height of only 0.58m[4], results derived from a study using the rather toy-sized Robosapiens RSMedia may only partially apply to more human-sized robots which, however, are more likely to be used in people's households and to regularly interact with humans in the future.

A further study investigating both the individual and combined contribution of gazing and co-verbal gesture to the persuasiveness of a story-telling robot was presented by Ham et al. (2011). In four experimental conditions, the NAO robot (see Figure 3.10b) told participants a persuasive story about the consequences of lying using the following communication channels: 1) speech only, 2) speech with gaze, 3) speech with gesture, or 4) speech with both gaze and gesture. The persuasive effect was measured by using post-experimental questionnaires asking participants to evaluate the lying individual from the robot's story. Results showed that gazing had a significant effect on the robot's persuasiveness, i.e., participants being gazed at by the robot were persuaded more by the message conveyed by the story than participants who were not gazed at. In addition, a robot using gestures was found to be more persuasive when it additionally used gazing than when it only gestured along with speech, suggesting that the combination of gesture and gaze can increase the persuasive effect of the robot. Similar to the study conducted by Kim et al. (2008), however, Ham et al.'s experiment lacked real interaction between humans and the robot, as participants were asked to only observe and listen to NAO while it was telling the story.

Also using the NAO robot, Park et al. (2011) studied the effects of robot gesture on social conversation in a simple HRI situation. In their experiment, gestural behavior was manipulated in two experimental conditions: gesture vs. no-gesture. Participants were first asked to observe the robot during a four-minute exposure in which NAO, depending on the condition, either used only speech or gesture along with speech. Subsequently, participants were invited to rate their perception of the robot based on several questionnaire items. In summary, results showed that the gesturing robot was perceived as having a higher level of conversation proficiency and appeared to be more familiar as well as human-like than the one using speech only. Furthermore, participants expressed a greater desire to talk to the robot that displayed speech-accompanying gestures than to the robot that was not employing such non-verbal behavior. As in the study presented by Ham

---

[4]http://www.wowwee.com/en/support/rs-media - accessed December 2011

et al. (2011), weaknesses of the experiment are the non-interactive character of the test scenario as well as the small size of the NAO robot in use.

In conclusion, the review of empirical findings resulting from the study of gesture-based HRI as presented in the relevant literature shows that the evaluation of such robot behavior is still in the early stages, thus offering ample scope for further research. At the same time, common limitations of previous studies, as described above, can be identified. First, due to their availability and affordability, small-sized robots such as NAO and Robosapiens RSMedia are increasingly used to study human-robot interaction. This may certainly lead to a number of empirical implications, however, it remains questionable whether the results obtained from studies using such toy-sized robots can be generalized, especially with regard to human-sized robots. Second, in line with the criticism outlined in Section 3.2.1 with regard to technical implementations, many studies focus on the evaluation of a single type of gesture, typically deictics. Finally, most evaluation studies are conducted in robotic laboratories which are not designed to provide an environment familiar to human participants. Although unavoidable and valid, it is likely that the unfamiliar atmosphere in addition to the unusual experience of interacting with a robot may influence the outcome of the studies.

3.3 Summary and Discussion

Putting the major objectives of the present work into context, the purpose of this chapter was to provide an overview of the current state of the art with regard to computational approaches to the generation and evaluation of speech and gesture for artificial communicators. Two research areas were identified as relevant, thus motivating the structure of the chapter: first, the area of virtual agents, in which systems for the realization and evaluation of multimodal communicative behavior for computer animated characters have been developed; second, the field of robotics, in which a variety of approaches to the generation and analysis of speech-accompanying non-verbal behaviors for social robots have been presented.

Accordingly, in **Section 3.1** the first point of focus, namely the review of contributions from the virtual agent community, was addressed. Initially, Section 3.1.1 offered a technical perspective by presenting and discussing the following virtual agent frameworks allowing for multimodal communicative behavior: *Animated Conversation*, *REA*, *BEAT*, *GRETA*, and finally, *MAX*, which provides the foundation for the technical work of this thesis. As introduced, a core part of the behavior generation system underlying the agent MAX is formed by the Articulated Communicator Engine (**ACE**). In contrast to the other presented approaches, ACE was outlined as being the first virtual agent framework to provide for mutual adaptation mechanisms between the timing of speech and gesture at run-time. This was discussed as being advantageous, since it can overcome several issues encountered in previous systems by providing greater flexibility and the possibility to generate gestures without predefined templates or detailed keyframe descriptions.

Moreover, inspired by theories from human gesture research, ACE is built on the assumption that the production of continuous speech and gesture is organized in successive coherent chunks. For behavior generation with ACE, multimodal utterances can optionally be specified as **keyframe animations** or as **feature-based descriptions** using the Multimodal Utterance Representation Markup Language (**MURML**). Finally, the innovative quality of the ACE framework manifests itself in an **incremental on-line scheduling** mechanism for multimodal agent behavior, which handles cross-modal interactions at different levels of an utterance. Due to these advanced features, ACE was identified as a suitable technical basis and starting-point for the development of a speech-gesture generation model for social robots as intended by the present research.

Complementing the technical view on virtual agent systems, Section 3.1.2 provided empirical findings from experimental studies conducted with such agents. More specifically, a selection of studies investigating the effect of hand and arm gestures on human perception of virtual agents, e.g., with regard to traits such as naturalness, believability, and likability, were reviewed. In summary, in most cases the agent's gesturing behavior was found to positively affect the way in which it was perceived and rated by human participants.

In **Section 3.2**, the second point of focus was then introduced and discussed by presenting a number of related robotic systems from both technical and empirical points of view. To begin with, Section 3.2.1 provided a categorization of characteristic features of related technical approaches to robot gesture generation and further illustrated them by outlining a selection of implementation examples. Generally, existing systems can be classified to belong to one or more of the following categories characterizing the given approach:

- **Predefined gestures and limited body expressiveness.** The system is equipped with a set of predefined or pre-recorded gestures which are not generated on-line during HRI. The robot's expressiveness may be limited due to less complex body design, e.g., with less DOF, no hands, restricted mobility, etc.

- **Unfavorable appearance of the gesturing robot.** The robot in use may expose only little or no humanoid traits, look 'uncanny', be too small, or may be otherwise limited in its appearance, e.g., due to its design (see above).

- **Limitation to a single gesture type.** The approach is limited to the implementation and evaluation of a single type of gesture, instead of providing a general framework able to handle all, or at least a variety, of gesture types.

- **Building on experience from the virtual agents domain.** Similar to the work presented in this thesis, the proposed system may build on elements taken from, and originally developed for, the virtual agents domain.

- **Open-loop control and unidirectional synchronization.** All previously presented approaches produce multimodal robot behavior using open-loop control, i.e., without sensory feedback, and unidirectional synchronization mechanisms, limiting the flexibility of the system and its adjustability at run-time.

In especially addressing the issue of open-loop control, the major technical contribution of the present work is thus to provide an implementation for co-verbal robot gesture generation which stands out from existing systems in multiple ways. First, the proposed approach uses an advanced humanoid and human-sized robot which is highly expressive due to its design, large number of DOF, and smooth body movements. Second, it builds on one of the most sophisticated virtual agent frameworks. Finally, and most importantly, for the first time in a robotic framework, it incorporates reactive closed-loop feedback for cross-modal synchronization (see RQ2).

Following this technical view on existing gesture generating robotic systems, Section 3.2.2 was dedicated to the empirical perspective on speech-gesture synthesis for communicative robots. Generally, building acceptable artificial communicators such as social robots demands a thorough understanding of the assumptions that humans make about such interactive machines and how they perceive them during interaction. To arrive at such an understanding of the dynamics of gesture-based HRI, it is therefore necessary to study real interactions between humans and social robots in suitable situations. However, only few relevant studies investigating the effects, perception, and acceptance of robot gesture in HRI could be found, with even less of them specifically focusing on hand and arm gesture in particular. Moreover, many common limitations were identified with regard to the experimental design or procedure of most studies, e.g., due to the lack of real interaction, the use of toy-sized robots, or the consideration of only one gesture type. Nevertheless, results generally suggest a beneficial effect of non-verbal robot behavior such as gesture on the human observer and interaction partner respectively. Since there is still ample scope and need for evaluation, the second major contribution of the present work is to provide a set of novel empirical findings about the effect that co-verbal communicative robot gesture may have on human interaction partners.

Part II: Technical Implementation

"Divide each difficulty into as many parts
as is feasible and necessary to resolve it."
René Descartes

Chapter 4

System Overview

Having provided background information from psycholinguistic (Chapter 2) and computational (Chapter 3) perspectives in the first part of the thesis, this chapter sets out to introduce the technical work of the present research. To endow a humanoid robot with communicative co-verbal gestures, a large degree of flexible control with regards to shape properties of the gesture is required. At the same time, adequate timing and natural appearance of these body movements are essential to add to the impression of the robot's liveliness. The challenges of generating flexible communicative robot gesture are highlighted by providing a general technical overview of the designated system.

First, in Section 4.1 inspiration is drawn from a neurobiological point of view according to which gesture generation can be viewed as a motor control problem to be solved by a hierarchical control scheme. This notion is readopted in Section 4.2 in which the existing modules, which constitute the starting points and the foundation for the technical implementation of the present work, are presented. Initially, the concept underlying gesture motor control in the multimodal behavior realizer ACE is reviewed. This is followed by a description of the Whole Body Motion (WBM) controller which provides a flexible kinematic framework to control upper body movement of the Honda humanoid robot. Finally, Section 4.3 provides an outline of the solution sought and a discussion of the main challenges faced when attempting to bridge the gap between ACE, the action generation framework for virtual agents, and WBM, the software controlling the humanoid robot. This way, the transition to the implementation work described in the subsequent two chapters is established.

4.1 Gesture Generation as a Motor Control Problem

From a neurobiological point of view, the generation of hand and arm gestures can be viewed as a specific, i.e., conceptual and communicative, form of movement of the upper limbs, serving a special purpose. At a more general level, human body movements can be classified into three distinct categories: *reflexive*, *rhythmic*, and *voluntary* (Ghez and Krakauer, 2000). Reflexive movements are involuntary, coordinated patterns of muscle contraction and relaxation in response to peripheral stimuli. Rhythmic motor movements are produced by repetitive patterns of muscle contraction and include swallowing, chewing, locomotion, and, of more relevance to the present research topic, beat gestures. Finally, and most importantly with regard to the work of this thesis, voluntary movements are goal-directed motions, i.e., they are initiated and performed to accomplish a specific task or goal. They are often referred to as 'motor skills' or just 'skills' and improve with practice as a result of feedback and feedforward mechanisms (Ghez and Krakauer, 2000).

Given the inherent communicative functions as well as the semantic and conceptual features that constrain their execution, gestural movements as considered in this thesis fall into this third category of **voluntary movements**. Indeed, from a more general point of view, gestures are comparable to other voluntary hand and arm motions in that they underly similar processes at the motor planning and control level. As a starting point for a technical analysis of gesture generation, this biologically focused viewpoint is borrowed and further elucidated in this section. Two core problems or challenges regarding human motor control have been identified as relevant to the present work, each addressing a central question:

- **Degrees of freedom problem.** How do particular forms of motor activity emerge when a multitude of movements potentially allow for the completion of the same task?

- **Hierarchical organization.** How can multiple limbs be centrally controlled and yet be simultaneously synchronized?

Focusing on the first of these two questions, the degrees of freedom problem which represents a central challenge of both human and artificial motor control is described and illustrated in Section 4.1.1. Subsequently, the hierarchical organization of motor control is highlighted and discussed in Section 4.1.2, thus addressing the second question as outlined above. Similar to the degrees of freedom problem, it has relevance in both biological and artificial kinematic systems. Although the following review is by no means exhaustive, it provides the basic understanding required to place the technical challenges of the implementation into a suitable context.

4.1.1 Degrees of Freedom Problem

Humanoid robots comprise a large number of degrees of freedom (DOF). Although this leads to an abundance of possible body movements for the robot, such increased flexibility results in a control problem. For example, when pointing to an object, many different postures can be adopted at the beginning, but in addition, each step on the way to the final posture and indeed the final posture itself may vary. Given a plethora of feasible motions for the same task, which particular movement should be chosen and performed by which motors and body parts? Equipped with even more DOF, the human body or, more specifically, the human central nervous system (CNS), is frequently faced with a similar problem. Consequently, the following question arises: which strategies does the human CNS use to select one specific movement from the infinite pool of possible solutions to solve the task at hand?

Although the underlying mechanisms have not yet been fully identified (Latash et al., 2007), research in fields such as Neuroscience and Neurobiology has provided a number of theories attempting to account for the human motor control problem. This problem belongs to the class of "ill-posed" problems because its solution is not uniquely determined (Kawato et al., 1990). This means, it cannot be solved without some additional information about the constraints imposed on the system. Since Bernstein (1967) was the first to draw attention to this kind of problem, it is often referred to as the *Bernstein problem* (e.g., Latash, 1993). It has also been named the *degrees of freedom problem* or the *problem of motor redundancy* (Saltzman, 1979; Turvey, 1990; Latash, 1996).

In setting out to solve the problem, the research field of robotics provides a suitable environment to develop sophisticated testbeds on which biologically inspired theories and hypotheses can be implemented and tested. The concepts that are most relevant to the technical work of this thesis are briefly described in the following.

4. SYSTEM OVERVIEW

Kinematics

Whenever goal-directed movements are generated, the degrees of freedom problem is encountered and must be solved. In the case of human motor control, the CNS has to convert spatio-temporal information about the target location into patterns of arm muscle activity in order to be able to move the arm towards the target (Stein, 2010). For example, for a pointing gesture towards an object, the target location of the tip of the index finger can be described with a three-dimensional vector X_T in external coordinates, while the start position of the finger tip can be described with another 3D vector X_S.

Even for this simple task of moving the hand from X_S to X_T, there are infinitely many possible paths along which the finger could move, and for each of these paths, there are infinitely many velocity profiles (i.e., trajectories) which the hand could follow. And even if the hand path and velocity have been specified, each point along the path can be accomplished by a magnitude of combinations of joint angles and arm configurations respectively. These, in turn, can be achieved by many different muscle activations (Jordan and Wolpert, 1999).

The problem of the motor control system is therefore to select one out of an infinite number of possible trajectories and to generate the chosen movement of the arm which moves the tip of the index finger from X_S to X_T. A schematic illustration of the problem is shown in **Figure 4.1**. Generally, such trajectory can be specified in two different ways according to how the configuration of the arm is represented for each point along the trajectory:

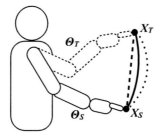

Figure 4.1: Schematic illustration of the degrees of freedom problem: to generate a movement from the initial location X_S to the target location X_T, the control system must determine a specific trajectory; three possible end-point trajectories are illustrated.

1) **Joint space.** At a non-dynamic level, the posture, or configuration, of the arm can be described by internal coordinates as a *joint level representation* Θ. Given the above mentioned example (see Figure 4.1), the space of joint angles for the start configuration of the arm and hand is described by the vector Θ_S which directly corresponds to the 3D point X_S. Establishing the configuration of the end-effector (in this example the finger tip) based on the relative configuration of each joint is referred to as **forward kinematics**.

2) **Task space.** Alternatively, a *task level representation* of the desired end-effector location, e.g., based on the external coordinates of X_T, can be used to determine a corresponding target joint angle configuration Θ_T. However, since there is no unique solution, it constitutes an ill-posed problem. Establishing the configuration of each joint based on the position of the end-effector is referred to as **inverse kinematics**.

Given these two different representation formats, coordinate transformations between intrinsic joint space and extrinsic task space are an essential part of motor control, closing the sensorimotor loop (Jordan and Wolpert, 1999; Stein, 2010).

4.1.2   Hierarchical Model of Motor Control

Empirical evidence suggests that the human CNS plans goal-directed movements in extrinsic task space coordinates rather than in joint space (e.g., Wolpert et al., 1995; Flanagan and Rao, 1995). This assumption is based on the *principle of smaller complexity of the representation* (Morasso, 1986), since the task-level representation of a target position X_T is lower-dimensional than the corresponding joint space description Θ_T. This view has been further supported by studies demonstrating that trajectories of a large class of unconstrained movements are characterized by several invariant spatio-temporal features at the task level, which cannot be found at the joint level (e.g., Morasso, 1981; Abend et al., 1982; Flash and Hogan, 1985; Gordon et al., 1994; Haggard et al., 1995). For example, a wide range of movements show a tendency towards roughly straight hand paths in external space with single-peak, bell-shaped velocity profiles.

These invariant properties have been partially explained by so-called **Generalized Motor Programs** (GMP; Schmidt, 1982) which refer to a set of general schemes associated with specific motion classes, e.g., for pointing movements. By suggesting that one motor program can be executed in a variety of different ways, thus comprising an entire class of movements, this notion has replaced the traditional *motor program* concept (Keele and Posner, 1968) which assumes a different motor program to be memorized for each movement.

Given an extrinsic task goal, the CNS needs to translate the target location X_T into neural commands that activate the muscles required to move the arm from its initial position to the final position. The computational process of **motor planning** is concerned with the selection of a single solution at each level of the motor control hierarchy from the many alternatives that comply with the task. As illustrated in **Figure 4.2**, at any level, one pattern of behavior corresponds to many patterns at the level below (*one-to-many*), but directly and uniquely specifies the pattern at the level above (*many-to-one*).

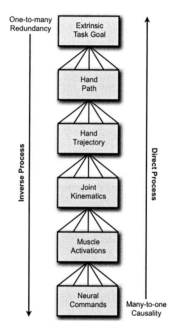

Figure 4.2: Hierarchical model of motor control adapted from Jordan and Wolpert (1999); specifying a pattern of behavior at any level complies with many patterns at the level below (*one-to-many*), but directly corresponds to the pattern at the level above (*many-to-one*).

Assuming that a target joint angle configuration Θ_T has been found and that the starting vector of the joint angles Θ_S is known, the next step in solving the control problem is to find patterns of joint torques that move the arm from Θ_S to Θ_T (Latash, 2008). This task is referred to as the problem of **inverse dynamics** (Hollerbach and Atkeson, 1987). Since the start and end configurations of the arm do not uniquely define the arm's trajectory, i.e., there are infinitely many time functions $\Theta(t)$ that can move the fingertip from X_S to X_T, the problem of inverse dynamics is another ill-posed problem. Once such patterns of joint torques $T(t)$ have been computed, the next task for the motor controller is to compute patterns of muscle force $F_M(t)$ that can produce $T(t)$. The CNS actively controls muscle forces by changing the levels of muscle activation $A_M(t)$. Finally, the last step of the problem is to compute physiological command signals $C(t)$ which the CNS will send to neurons in the spinal cord to activate α-motoneurons. These, in turn, activate the involved muscles and, as a result, the movement is generated (Latash, 2008). In a simplified way, the whole process of computing the movement can be represented as:

$$\{X_T\} \to \{\Theta_T\} \to \{T(t)\} \to \{F_M(t)\} \to \{A_M(t)\} \to \{C(t)\} \qquad (4.1)$$

Once the computation is completed, control signals $C(t)$ are sent to spinal neurons which modify the activation levels $A_M(t)$ of according muscles. As a result, the muscles produce the force patterns $F_M(t)$ required to rotate the joints from their initial to their final configuration by means of joint torque patterns $T(t)$. Finally, the target configuration Θ_T and location X_T respectively will be achieved, and the task will be accomplished:

$$\{C(t)\} \to \{A_M(t)\} \to \{F_M(t)\} \to \{T(t)\} \to \{\Theta_T\} \to \{X_T\} \qquad (4.2)$$

In reality, this process underlies several non-trivial computational steps. There is an ongoing debate as to how the CNS computes a transformation of the motor plan into actual motor commands. Several models have been proposed in the literature, with the **force-control approach** and the **equilibrium point hypothesis** forming the two most prominent concepts (Latash, 2008; Stein, 2010).

In line with the above mentioned concept of Generalized Motor Programs, the force-control approach assumes that the CNS generates command signals based on pre-computed patterns of muscle forces that are appropriate to accomplish the given task. This view further supports the notion of *internal models* which mimic or simulate the behavior of the controlled motor system within the CNS (Wolpert, 1997; Jordan and Wolpert, 1999). According to this view, it is assumed that the CNS incorporates two types of models. First, **inverse models** compute neural commands that are required to achieve the desired mechanical effect, i.e., they model sensory-to-motor transformations (Equation 4.1). Second, **direct** or **forward models** predict the mechanical effects of current neural commands on the (next) state of the system, i.e., they model motor-to-sensory transformations (Equation 4.2).

An alternative view based on the equilibrium point hypothesis, which was first introduced by Feldman (1986), suggests that the use of innate reflex patterns can significantly contribute to the solution of the degrees of freedom problem. According to this hypothesis, movements are not explicitly programmed, but result from dynamical properties of the musculoskeletal system itself, i.e., from shifts in the equilibrium position of the muscles. More specifically, thresholds for muscle stretch reflexes are centrally modified, causing the limb to be out of equilibrium until the CNS establishes the new equilibrium position (Rosenbaum, 2002). Based on these mechanisms, posture stabilization is turned into movement production (Latash, 2008).

In an attempt to reconcile the two above described approaches, Latash (1993) proposed a hierarchical motor control scheme as illustrated in **Figure 4.3**. It comprises three major steps of motor control:

1) **Internal simulation.** The movement is first planned in terms of its extrinsic end-point trajectory leading from the initial hand location X_S to the target position X_T. To this end, an internal simulation of the planned movement is performed, which accounts for predictable changes in the conditions of movement execution and results in a function reflecting the kinematic properties of the desired trajectory.

2) **Generation of motor command.** The simulated trajectory is translated into motor command variables which can be interpreted by the lower-level structures of the system.

3) **Execution.** The specified motor commands are executed, resulting in altered activation levels of the relevant muscles and, finally, to the generation of a movement that ideally matches the simulated trajectory.

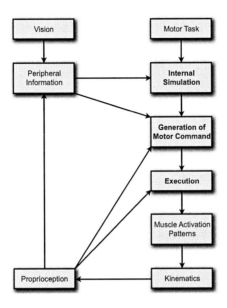

Figure 4.3: General scheme of motor control adapted from Latash (1993) comprising three major steps: internal simulation, generation of motor command, and execution.

4.2 Existing Modules

One of the main characteristics of the present approach to the technical realization of communicative robot gesture is the reuse of an existing solution from the domain of virtual agents. This approach is not only beneficial in that it circumvents the need to build a new framework for robot gesture from scratch; it also serves as a proof of concept showing that within the scope of gesture realization, transfer from virtual character animation to robot behavior generation is indeed feasible.

Two major modules have been available for use prior to the technical realization of the present project: first, the multimodal behavior generation framework of the Articulated Communicator Engine (ACE) which has been previously employed only for the animation of embodied conversational agents; second, the Whole Body Motion (WBM) controller which provides a flexible kinematic framework for controlling upper body movement of the humanoid robot. Together, ACE and the

83

robot's WBM software represent the pre-existing modules that form the starting point and foundation for the technical implementation of the present work. They are briefly described in Sections 4.2.1 and 4.2.2 respectively, with a special focus on those aspects of motor control that are relevant to co-verbal gesture generation.

4.2.1 Articulated Communicator Engine (ACE)

Some general technical features of the ACE framework, especially with regard to its synchronization mechanism, have already been highlighted in Section 3.1.1. This section sets out to elucidate the system from a motor control point of view by illustrating how arm gestures are generated in ACE for virtual character animation. Generally, gesture production in ACE comprises three main stages as visualized in **Figure 4.4** and described in the following.

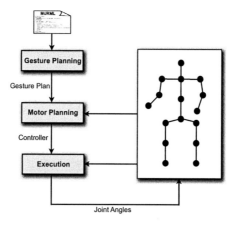

Figure 4.4: Gesture production stages in ACE (adapted from Kopp, 2003).

First, during high-level **gesture planning** the expressive phase of a gesture, i.e., the stroke, is defined by a set of movement constraints as formulated in the XML-based Multimodal Utterance Representation Markup Language (MURML; Kranstedt et al., 2002). At this stage, the corresponding body parts are allocated and the timing of the stroke phase is determined. Second, at the lower-level **motor planning** stage, a solution to the motor control problem posed by the previously resolved specification and the resulting movement constraints is sought. Based on a kinematic model of human hand and arm movement, a motion sequence which satisfies the given spatio-temporal movement constraints is determined. Finally, at the **execution** stage, the selected movement sequence is translated into action, i.e., the planned behaviors are executed by the animated character.

In light of the present work, aspects and features of the ACE system which are relevant to the first two above described gesture production stages are briefly explained in the following. A more detailed description of the intricacies of the original ACE implementation can be found in Kopp (2003) and Kopp and Wachsmuth (2004).

Behavior Description in MURML: Basic Elements and Attributes

As shown in Figure 3.5 of the previous chapter, any behavior execution in ACE is initiated by an utterance description specified in MURML. Accordingly, a feature-based MURML file as illustrated in Figure 3.6 contains all the relevant information and constraints specifying the spatio-temporal characteristics of the utterance to be generated. It can thus be considered the description of an extrinsic task goal forming the beginning of a motor control task (cf. Figures 4.2 and 4.3). In the following, an explanation of how the task goal is encoded in MURML and how the given elements and attributes are interpreted by the ACE system is given; it is based on a selection of basic components.

As illustrated by the example file shown in **Figure 4.5**, every MURML description specifies exactly one utterance. Each utterance contains a maximum of one locution and one gesture unit as described in Section 2.2.1 and illustrated in Figure 2.5. Each utterance description begins and ends with the definition root tag element (<definition> ... </definition>) within which a nested tag element defines the beginning and end of the utterance (<utterance> ... </utterance>). A spoken utterance is defined using the <specification> tag and can be further annotated with time identifier tags to allow for the allocation of affiliated gestures. If the utterance contains more than one chunk, the transition between each two chunks is marked by the additional attribute chunkborder="true" within the <time/> tag.

In the case of a multimodal utterance, the speech specification is followed by the description of non-verbal behaviors marked by at least one <behaviorspec> tag which is assigned a unique identifier attribute (e.g., "gesture_1" and "gesture_2" in the example file of Figure 4.5). Within this behavior specification tag, a gesture is

```
<definition><utterance>

    <specification>
    You can take <time id="t1"/> this chair <time id="t2" chunkborder="true"/> and sit down on it. <time id="t3"/>
    </specification>

    <behaviorspec id="gesture_1">
        <gesture>
            <affiliate onset="t1" end="t2"/>
            <constraints>
                <parallel>
                    <static slot="HandShape" value="BSffinger"/>
                    <static slot="ExtFingerOrientation" value="DirA"/>
                    <static slot="PalmOrientation" value="DirL"/>
                    <static slot="HandLocation" value="LocChest LocRight LocNorm"/>
                </parallel>
            </constraints>
        </gesture>
    </behaviorspec>

    <behaviorspec id="gesture_2">
        <gesture>
            <affiliate onset="t2" end="t3"/>
            <constraints>
                <symmetrical dominant="left_arm" symmetry="SymMS">
                    <parallel>
                        <static slot="HandShape" value="BSflat"/>
                        <static slot="ExtFingerOrientation" value="DirA"/>
                        <static slot="PalmOrientation" value="DirD"/>
                        <dynamic slot="HandLocation">
                            <dynamicElement type="linear">
                                <value type="start" name="LocShoulder LocPeripheryLeft LocNorm"/>
                                <value type="end" name="LocAbdomen LocPeripheryLeft LocNorm"/>
                            </dynamicElement>
                        </dynamic>
                    </parallel>
                </symmetrical>
            </constraints>
        </gesture>
    </behaviorspec>

</utterance></definition>
```

Figure 4.5: A feature-based MURML specification for a multimodal utterance.

defined by the <gesture> tag which, in turn, contains further elements describing the properties of the gesture. First, the speech affiliate of the gesture is referenced within the <affiliate> tag by stating the according time identifiers to define the onset and end of the affiliate. Thus, it encodes the time constraints imposed on the gesture to allow for the desired multimodal synchronization (cf. Section 3.1.1).

Second, the form features of the gesture are specified within the <constraints> element node. Generally, a gesture can be performed either with one hand (by default; see "gesture_1") or with two hands (see "gesture_2"). Two-handed gestures are specified by an additional <symmetrical> tag with further attributes determining the dominant arm and the symmetrical relationship between the two arms. Furthermore, movement constraints can be specified in a way that they either apply simultaneously (marked by a <parallel> tag) or successively (alternatively marked by a <sequence> tag).

Finally, a gesture can either have a static stroke, i.e., forming a *stroke hold* (cf. Section 2.1.3; see "gesture_1"), or a dynamic stroke, e.g., by moving the hand from a start position to an end location (see "gesture_2"). A static gesture is annotated by <static> tags including attributes about the according features (slot) and their assigned values (value). In contrast, a dynamic gesture contains at least one feature (slot) that is embedded in a <dynamic> tag. This, in turn, comprises a <dynamicElement> tag with two value attributes specifying the start (<value type="start"...>) and end (<value type="end"...>) configurations respectively of the according hand feature. There are four slot types to specify the overt form of the designated gesture stroke in a conceptual fashion:

1) "HandShape" describes the shape of the gesturing hand, which can either be an open flat hand (BSflat), a clenched fist (BSfist), or a flat hand with splayed fingers (BSffinger). Additional values can be used to further specify the configuration of individual fingers and the thumb respectively.

2) "ExtFingerOrientation" specifies the extended finger orientation along the back of the hand, which can either be directed to the front (DirA), to the back (DirT), to the left (DirL) or right side (DirR), or upwards (DirU) or downwards (DirD). Combined values are possible, e.g., DirAL for front left.

3) "PalmOrientation" determines the orientation of the palm which can be defined by the same absolute values as used for the extended finger orientation. Together with the ExtFingerOrientation slot, PalmOrientation specifies the wrist orientation of the hand.

4) "HandLocation" defines the relative location of the wrist based on the division of human gesture space as proposed by McNeill (1992), which is illustrated in **Figure 4.6**. Three positional dimensions can be specified: the longitudinal position (e.g., LocAboveHead, LocChest, or LocAbdomen), the transversal position (LocCCenter by default or, e.g., LocRight or LocPeripheryLeft), and the sagittal position (LocNorm by default or, e.g., LocNear or LocFar).

These abstract symbols specifying the gesture form are translated into a formal movement plan by the gesture planner and, eventually, into concrete command values by the motor planner of the ACE system (see Figure 4.4). A detailed description of elements, attributes, and possible values for gesture description in MURML can be found in Kranstedt et al. (2002) and Kopp (2003).

87

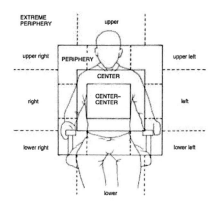

Figure 4.6: Division of human gesture space according to McNeill (1992, p. 89).

Kinematic Body Model

As one of the first demonstration applications developed for the ACE behavior generation toolkit, the virtual agent MAX (cf. Section 3.1.1) is used as an example implementation in this thesis to illustrate the capabilities of the underlying system.

To allow for the interpretation and appropriate execution of the spatio-temporal features encoded in a given MURML specification, ACE is tightly coupled with the kinematic body model of the embodied agent to be animated. An anthropomorphic kinematic skeleton was thus defined for MAX (see **Figure 4.7a**), with a total of 68 segments and 103 degrees of freedom (DOF) in 57 joints. These include 53 DOF in 25 joints for the body (see **Figure 4.7b**) and 25 DOF in 16 joints for the fingers of each hand (see **Figure 4.7c**). The kinematic skeleton has a height of 182 cm and is subject to realistic joint limits based on the Standard Humanoid Animation (H-Anim) body model[1]; all properties are specified in an XML file.

The MAX body model represents the initial skeleton definition included in the ACE system used, serving as a pre-configured starting point for the transfer of the system to a robotic platform as intended by the present work. The model and its complete kinematic properties are described in more detail in Kopp (2003).

---

[1]Version 1.1; http://h-anim.org/Specifications/H-Anim1.1/ – accessed February 2012

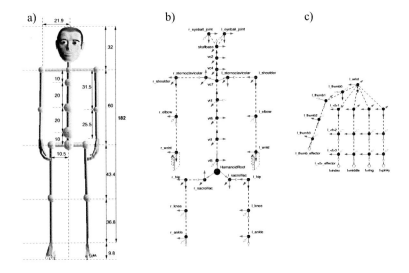

Figure 4.7: Kinematic body and hand models of the virtual agent MAX (reprinted from Kopp, 2003, p. 100ff.): a) dimensions (in cm) of the kinematic body model; b) segments, DOF, and joints of the kinematic skeleton; c) segments, DOF, and joints of the kinematic hand.

Motor Control in ACE

Inspired by biological models as described in Section 4.1, gesture motor control in ACE is realized hierarchically. During higher-level planning, the motor planner is provided with timed form features of the required gesture as annotated in the MURML specification. To solve the given motor control problem, this information is then passed on to independent motor control modules. The idea behind this functional-anatomical decomposition of motor control is to break down the complex control problem into solvable sub-problems (Zeltzer, 1982).

Accordingly, the motor planner in ACE provides specific modules, among others, for the arms, the wrists, and the hands (see **Figure 4.8**). These modules, in turn, instantiate **local motor programs** (LMPs) which are used to animate required sub-movements. LMPs operate within a limited set of DOFs and over a designated period of time, e.g., during the preparation, stroke, or retraction phase

89

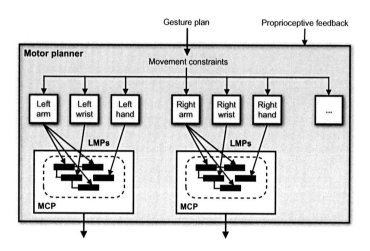

Figure 4.8: Composition of the ACE motor planner (adapted from Kopp and Wachsmuth, 2004).

of the gesture (see **Figure 4.9**). For the motion of each limb, an abstract **motor control program** (MCP) coordinates and synchronizes the concurrently running LMPs, gearing towards an overall solution to the control problem. The top-level control of the ACE system, however, does not attend to how such sub-movements are controlled.

To ensure an effective interplay of the LMPs involved in a MCP, the planning modules arrange them into a controller network which defines their potential interdependencies for mutual (de-)activation. LMPs are able to transfer activation between themselves and their predecessors or successors to allow for context-dependent gesture transitions. Thus, they can activate or deactivate themselves at run-time depending on feedback information on current movement conditions. Once activated, LMPs are continuously applied to the kinematic skeleton in a feedforward manner; this process is coordinated by their respective MCPs.

Since gesture generation is based on external form features as given in the MURML description, arm movement trajectories are specified directly in task space. For each animation frame, externally formulated LMPs for wrist position, preparation and stroke of wrist flexion, and swivel movement are invoked first.

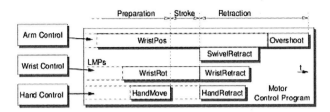

Figure 4.9: Composition of motion control in Local Motor Programs (LMPs) for a hand-arm gesture (adapted from Kopp and Wachsmuth, 2002).

Subsequently, the inverse kinematics of the 7-DOF anthropomorphic arm is solved using an analytical algorithm from the IKAN package[2], which offers a complete set of real-time inverse kinematics algorithms suitable for anthropomorphic limbs such as arms or legs (Tolani et al., 2000). Details of the exact algorithms used for trajectory formation and to solve redundancies of the arm can be found in Kopp (2003) and Kopp and Wachsmuth (2004).

To account for temporal constraints imposed by the concurrent speech modality, on-line timing of gestures is accomplished as follows. As highlighted in Section 3.1.1, synchrony within a chunk is generally achieved by adapting the gesture to structure and timing of speech. To do this, the ACE scheduler retrieves timing information about the synthetic speech at the millisecond level and defines the start and the end of the gesture stroke accordingly. Such temporal constraints, e.g., specifying how long the hand has to form a certain shape, are automatically propagated down to each single gesture component. The motor planner then creates the LMPs that meet the specified temporal constraints in addition to the form constraints of the gesture.

The second aspect of scheduling, namely the decision to skip preparation or retraction phases, results from the interplay of motor programs at run-time. Motor programs monitor the current movement state of the body and are autonomously activated to realize the planned gesture stroke as scheduled. Whenever the motor program of the following gesture takes over the control of the effectors from the preceding program, the retraction phase turns into a transition into the next

---

[2]http://cg.cis.upenn.edu/hms/software/ikan/ikan.html/ – accessed February 2012

gesture. Such on-line scheduling is made possible by the interleaved production of successive chunks and leads to fluent and continuous multimodal behaviors.

4.2.2 Whole Body Motion (WBM)

Humanoid robots have a complex structure, many DOF and multiple end-effectors. Controlling the whole body of the robot is a complex task due to the high dimensionality and redundancy of the system. Thus, the specification of motions based on lower-dimensional end-effector tasks is desirable for controlling the robot.

Besides the ACE system, the second pre-existing module available for the technical implementation of the present work is the Whole Body Motion (WBM) controller of the humanoid robot. This provides a kinematic framework which allows for flexible real-time control and generation of upper body movements of a redundant robot. Specifically, the WBM module aims at controlling all DOF of the humanoid robot by given end-effector targets. Relevant aspects and features of the framework are briefly described in the following. More detailed information on the WBM system can be found in Gienger et al. (2005, 2006).

Kinematic Body Model

The humanoid robot used for the work of this thesis has a height of 120 cm and a weight of 52 kg;[3] the kinematic model underlying the WBM controller is illustrated in **Figure 4.10** (Gienger et al., 2005). In the initial configuration, the x-axis points forward, the y-axis points to the left, and the z-axis points upward. Roll, tilt, and pan specify a rotation about the x-, y-, and z-axis respectively.

The first link corresponds to the heel coordinate system which is centered between the feet and aligned with the heel edge. It comprises 3DOF in the form of translations in the x- and y-direction, and a rotation about the z-axis. The following links correspond to the body segments of the robot. The pelvis is subject to three translations and rotations with respect to the heel frame. The head is connected to the upper body and comprises 2DOF composed of pan and tilt joints. The kinematic model further includes two arms comprising 5DOF each, with 3DOF in the shoulder, 1DOF in the elbow, and 1DOF in the wrist allowing

---

[3]http://asimo.honda.com/downloads/pdf/honda-asimo-robot-fact-sheet.pdf – accessed February 2012

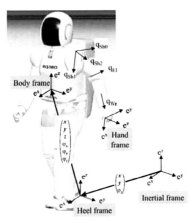

Figure 4.10: Kinematic body model of the Honda humanoid robot (reprinted from Gienger et al., 2005).

for a rotation of the hand with respect to the forearm. Each hand is represented by an additional coordinate system defining a hand reference point for a 1DOF grasp axis. In total, the model comprises 21 DOF.

Shifting and rotating the pelvis results in a one-to-one mapping onto the leg joints, which is implemented within a separate leg and balance controller as an independent process. As a result, the leg joints are not explicitly included in the model, but are accounted for by the DOF of the upper body instead. The state vector consists of the DOF that can be directly or indirectly controlled and thus comprises

$$
q = \begin{pmatrix}
\left(_I x_{hl} \; _I y_{hl} \; _I \varphi_{z,hl} \right)^T \\
\left(_{hl} x_{ub} \; _{hl} y_{ub} \; _{hl} z_{ub} \right)^T \\
\left(_{hl} \varphi_{x,ub} \; _{hl} \varphi_{y,ub} \; _{hl} \varphi_{z,ub} \right)^T \\
\varphi_{arm,L}^T \\
\varphi_{arm,R}^T \\
\left(\varphi_{pan} \; \varphi_{tilt} \right)^T
\end{pmatrix}
\tag{4.3}
$$

with indices I, hl, and ub denoting the inertial, heel, and upper body frames respectively (Gienger et al., 2006).

Motion Control with WBM

The WBM system allows for flexible kinematic motion control of a redundant robot by employing a control scheme that solves the redundant inverse kinematics problem on the velocity level. The underlying control algorithm is based on a tree-like kinematic description of the robot as illustrated in **Figure 4.11**, in which each leg and arm as well as the head are represented by a kinematic chain rooted in the upper body. The individual links within this hierarchy are either connected via joints and their respective DOF or via fixed rigid body transformations (Gienger et al., 2010). It is further possible to specify additional information for the DOF, such as a range or velocity limits.

For the calculation of forward kinematics, the transformations of all tree nodes are computed based on the current configuration of the robot (i.e., its joint angles), which corresponds to a descending tree traversal starting from the root node. For the computation of inverse kinematics, the "redundancy resolution" method first introduced by Liégeois (1977) is applied. This divides the control objective into a task and null space: the task space trajectory is projected into the joint space using a weighted generalized pseudo-inverse of the task Jacobian, while the null space is exploited to account for joint limit avoidance. More specifically,

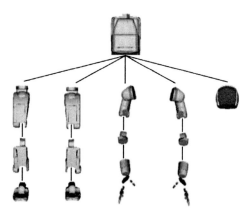

Figure 4.11: Kinematic tree structure underlying the motion controller of the humanoid robot; the kinematic chain of each limb is rooted in the upper body (adapted from Gienger et al., 2010).

redundancies are resolved by mapping the gradient of a given joint limit avoidance criterion into the null space of the motion (Gienger et al., 2010).

In addition, this process is optimized in compliance with a real-time collision avoidance algorithm as described in Sugiura et al. (2006, 2007) which protects the robot from self-collisions. For this, avoidance movements are blended with the WBM control by changing the priority between target reaching and collision avoidance depending on the current configuration of the robot. The WBM controller is further coupled with a separate walking and balancing controller which stabilizes the motion (Hirose et al., 2001). Since WBM may lead to displacements of the center of gravity, the balance controller can react by generating an upper body shift. This, in turn, is tracked by and incorporated into the WBM system without the need for active control.

As a major advantage of WBM control, motion targets may be specified selectively, e.g., joint angles of particular joints may be specified in addition to the desired task space trajectory. Furthermore, the task space trajectory may be defined so as to include displacement intervals which allow for the specification of a valid region around the target trajectory for smoother motion generation. Either way, remaining redundancies are solved based on the above described optimization criteria, resulting in flexible and robust motion control.

The formalized mathematical model and a more detailed description of the velocity-based WBM algorithm can be found in Gienger et al. (2005, 2006, 2010). The incorporated collision avoidance algorithm is further described and formalized in Sugiura et al. (2006, 2007).

4.3 Bridging the Gap - the Solution Sought

Given the pre-existing software modules ACE and WBM control, an important research question centers on the main challenges faced when attempting to bridge the gap between these two systems from different application domains (see RQ3). In order to endow the humanoid robot with co-verbal gestures while standing to benefit from the speech-gesture production model in ACE, an exploration and evaluation of the reusability of the existing features is necessary. In the following, an outline of the desired system architecture is presented and discussed and the challenges faced at different levels of the behavior generation process are elucidated.

4.3.1 Outline of the System Architecture

The development of an ACE-based action generation framework for the WBM controlled humanoid robot as intended by the present work targets a maximum exploitation of the already existing systems. On the one hand, ACE provides modules for the on-demand planning and synthesis of gestural motor actions and verbal utterances. These are coupled with an incremental process model to schedule the output behaviors into synchronized fluent utterances. On the other hand, the humanoid robot is equipped with the previously described WBM controller, which allows for the selective specification of motion targets both based on forward and inverse kinematics. Based on the features of the two underlying modules, the system architecture for a multimodal ACE-based action generation framework for the humanoid robot is outlined in **Figure 4.12** and further described in the following.

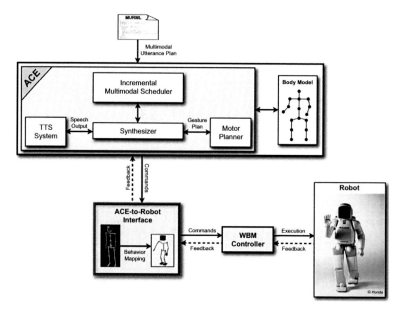

Figure 4.12: Outline of the system architecture for an ACE-based multimodal action generation framework for the humanoid robot; the required interface between the ACE system and the WBM controller is framed in red.

The overall system forms a generation pipeline integrating ACE with a body model specification at the beginning of the process and the WBM controller with the robot at the end of it. In between the two systems, a bi-directional interface – using both efferent control signals and afferent sensory feedback – connects the behavior planning and scheduling layer on the one side with the robot motor control layer on the other side. The interface is situated at the original animation level of the ACE framework, thus replacing the animation of the virtual agent with the behavior mapping to the robot and the coupling with the WBM controller. The WBM controller, in turn, subsequently translates the issued commands into motor actions for execution on the robot. For this purpose, the ACE-to-robot interface has to convert motor commands for speech and gesture generated for the virtual agent into control variables that can be applied to the robot at run-time.

This process is explored and discussed in the following parts of this thesis. More specifically, the challenges faced at this stage of the generation process are first introduced in the next two sections. Generally, to enable the humanoid robot to flexibly produce speech and co-verbal gesture at run-time using the ACE framework, challenges are encountered at two different levels: first, at the motion generation level, and second, at the speech-gesture synchronization level. These are discussed in Sections 4.3.2 and 4.3.3 respectively. The actual implementation details and decisions made at both levels are presented in Chapters 5 and 6.

4.3.2 Challenges at the Gesture Generation Level

The generation of communicative gesture for an artificial agent such as a virtual human or humanoid robot represents a specific type of movement generation. Due to timing constraints imposed by accompanying speech, motion generation for gesture is more restricted than for arbitrary movements. Apart from such timing constraints, requirements with respect to defined form features of the gesture further restrict the desired movement: if conceptual information is to be expressed, for example, as conveyed by iconic gestures, the hand trajectory and potentially specific finger movements are essential in conveying the meaning of the gesture. In contrast, other goal-directed movements such as grasping typically provide more flexibility: the end-effector trajectory towards the target position can often be selected from a broader range of valid paths, all of which may ensure the successful accomplishment of the task.

At a more basic level, however, beyond the above mentioned timing and form feature issues, the generation of robot gesture as intended by the present work can be considered as a **motion retargeting problem** (Gleicher, 1998). That is, movements generated by or for one body with certain kinematic properties are to be transferred to another body with different kinematic characteristics. This problem is typically encountered in imitation learning scenarios in which a human demonstrator performs a motor action that is to be replicated and learnt by an embodied artificial agent (Billard, 2001; Schaal et al., 2003). In the present case, the original motion stems from the body model of an animated character with human-like kinematic properties (see Section 4.2.1), while the target body is that of a kinematically and physically more constrained humanoid robot.

The motion retargeting problem is associated with the so-called **correspondence problem** which refers to the mismatch between the different embodiments of the demonstrator and the imitator (Nehaniv and Dautenhahn, 2001, 2002). For example, a human tutor and an imitating humanoid robot may differ in arm length and in the range of motion provided by each joint. The required correspondence can be established at different levels, depending on the objectives of the imitation process and the task respectively. In general, two approaches to solving the correspondence problem can be distinguished (Schaal et al., 2003):

1) **Correspondence in external task space.** This approach represents a simplified solution to the problem, since external or task coordinates are mostly independent of the kinematic and dynamic properties of the demonstrator's body. The only transformation required for this process is a mapping from the demonstrator's body-centered external space to the imitator's body-centered external space, which is a linear transformation. If imitation is only required at task-level, i.e., if only the end-effector trajectory of the imitated movement is important to fulfill the given task, then this approach offers a straightforward and sufficient solution. For example, if the task is to draw a circle with the finger in the air, then a mapping of the task space trajectory will suffice to successfully accomplish the task.

2) **Correspondence in internal joint space.** Solving the problem in internal space is a more complex endeavor. If the demonstrator and the imitator have dissimilar bodies, joint angle trajectories of the demonstrator need to be shifted and scaled to match the segment lengths and range of motion provided by the imitator's body. Therefore, the movement can only be imitated approximately by reproducing only the most important sub-states of the motion. Under these

circumstances, the correspondence problem consists of the identification and definition of the movement features that need to be reproduced to fulfill the given task. Despite its complexity, this approach typically leads to a more accurate imitation. For example, if the task is to pantomime the wing flapping movement of a chicken by moving the bent elbows up and down, matching the end-effector position alone would not convey the correct meaning of the gesture; instead, shoulder and elbow joint angles represent important sub-states of this motion which need to be included in the mapping process.

Given these two options, the pros and cons of each approach have to be considered in view of the research objective of the present work. The decision made and reasons for the chosen approach with regard to the implementation of an ACE-based generation framework for robot gesture are discussed in Chapter 5.

Besides the motion retargeting and correspondence problem which result from the general difficulty of mapping motions from one body to another with less DOF, further aspects of the present approach add to the given challenge. For example, the body of the robot is subject to additional physical constraints which are not imposed on a virtual animated character. To illustrate, the robot is controlled by actual motors with various motor states, its motion underlies fixed velocity limits, and it must strictly avoid collisions with its own body or entities in the environment. However, in light of ACE being originally designed for a virtual rather than physical platform, it fails to adequately account for such physical properties and limitations. For this reason, these challenges must be explicitly addressed when transferring the ACE framework to the humanoid robot.

Another issue demanding consideration during implementation is how to choose an adequate transfer and mapping rate between the virtual agent framework and the robotic platform. For successful integration of the two existing systems into one combined framework, the required interface needs to synchronize the two competing sample rates of the ACE framework and of the WBM software controlling the robot. This issue is further elucidated in Chapter 5.

Finally, on a less technical level, a desired degree of smoothness as well as perceived naturalness of the generated robot gesture constitutes another challenge. Since the WBM controller is coupled with a balance controller, upper body motion of the robot can lead to compensating shifts and movements around the joint connecting the upper body with the legs. Such motion may be perceived as "hip" movements which are not typically performed by humans when gesturing. If it is perceived to significantly distort the appearance of the generated gestures, ways to limit these incidental movements may have to be found.

4.3.3 Challenges at the Speech Synchronization Level

As highlighted at the beginning of Section 4.3.2, the generation of co-verbal gesture movements is particularly constrained by the need for adequate temporal synchronization with speech. To ensure the desired co-expressive synchrony between the two modalities, the gesture stroke onset must precede or, at the latest, begin right at the onset of the nucleus of the speech affiliate (cf. Section 2.2.1).

Given the physical limitations imposed by the robot's body as highlighted in the previous section, movement adaptation to the temporal constraints of speech is more difficult to realize with a robot than with an animated virtual character. For example, due to physical joint velocity limits, the robot may not be able to perform the designated arm trajectory in the time scheduled and proposed by the ACE framework. This may potentially result in mistimed synchronization with the speech affiliate, e.g., if the gesture stroke is performed *after* the affiliate.

In contrast, when animating embodied computer characters, motion speed is not subject to such "hard" restrictions, since the velocity profiles applied for animation are much more flexible. For this reason, it is possible to realize speech-gesture synchronization for virtual agents merely based on gesture adapting to the timing of speech within a chunk. However, using a robotic platform characterized by the above mentioned limitations, this approach to cross-modal adaptation proves problematic. Synchronizing robot gesture with speech poses challenges at a completely different level than speech-gesture coordination in virtual agents. These challenges can be broken down into two problems as expressed by the following questions. First, if planning and scheduling of gesture movement as done by ACE do not result in successful synchronization of the robot's speech and gesture, how can a more reliable prediction of gesture execution time be obtained? Second, how can planning and scheduling errors be accounted for during execution, e.g., if the actual motion time exceeds the predicted value and would thus lead to speech-gesture mistiming?

To tackle these challenges arising at the speech-gesture synchronization level, the present work will provide a solution which accounts for the robot-specific limitations in order to achieve cross-modal synchrony in a customized and optimized way.

4.4 Summary and Discussion

This chapter introduced the first major objective of the present work, i.e., the technical implementation of an ACE-based behavior generation framework for a humanoid robot. To contribute to a general understanding of the challenges of producing flexible communicative robot gesture at run-time, an overview of the intended technical system was given.

Initially, in **Section 4.1**, a neurobiological perspective was adopted, showing how gesture generation can be viewed as a motor control problem. This notion was illustrated with the **degrees of freedom problem** in Section 4.1.1, which investigates how particular forms of motor activity emerge from a multitude of movements all of which allow for the completion of the same task. In this context, basic concepts of motion generation were introduced: first, **forward kinematics** which can be employed to establish the position of an end-effector based on its configuration description in **joint space**, and second, **inverse kinematics** which refers to the derivation of a joint configuration based on the end-effector location specified in **task space**.

In Section 4.1.2 the **hierarchical organization of motor control** was highlighted, suggesting that the human CNS plans goal-directed movements in the lower-dimensional task space rather than in intrinsic joint space coordinates. The CNS thus needs to translate the target location into neural commands that activate the muscles required to move the arm from its current position to the final position. Selecting a single solution at each level of the motor control hierarchy from the many alternatives that comply with the task is referred to as **motor planning**.

The notion of biologically inspired motor control was readopted in **Section 4.2** to present the pre-existing modules which constitute the starting points and the foundation for the technical implementation of the present work. First, gesture motor control in the multimodal behavior realizer ACE was illustrated in Section 4.2.1. For this, basic elements and attributes which can be used to specify multimodal behaviors in MURML were summarized, and an outline of the kinematic body model of the virtual agent MAX was provided. Furthermore, hierarchically organized motor control in ACE was depicted to introduce the concepts of **local motor programs** (LMPs) and **motor control programs** (MCP) as sub-components of the motor planner.

101

This was followed by a description of the Whole Body Motion (WBM) controller, i.e., the software that controls the humanoid robot, and of the kinematic properties of the robot's body in Section 4.2.2. The WBM software provides a flexible method to control upper body movement by only specifying relevant task dimensions selectively in real-time. Redundancies are solved on the velocity level and are optimized with regard to joint limit avoidance and self-collision avoidance, resulting in smooth and natural movement.

Finally, **Section 4.3** provided an overview of how the gap can be closed between behavior planning and scheduling in ACE on the one hand and generation of these behaviors on the robot via WBM on the other. The designated system needs to combine conceptual representation and planning provided by ACE with motor control primitives for speech and arm movements for the robot. The architecture proposed in Section 4.3.1 forms a generation pipeline connecting ACE with the WBM controller via a bi-directional interface for efferent and afferent signaling.

This approach was further elucidated with special focus on the challenges faced at different levels of the behavior generation process. First, difficulties encountered at the gesture generation level were highlighted in Section 4.3.2. In this regard, the **correspondence problem** was introduced, relating to the mismatch between the different embodiments of the virtual agent and the physical robot. Second, in Section 4.3.3 issues that may arise at the speech synchronization level were briefly highlighted and discussed. In particular, due to stricter velocity limits imposed on the robot's body, the potential of mistimed synchronization between gesture stroke and speech affiliate was pointed out. These challenges have to be tackled for successful realization of the technical objective at hand.

The presented outline of the solution sought and envisaged challenges of robot gesture generation provides a preview on the actual realization of the designated framework. Thus, this chapter forms a transition from the technical foundation to the implementation work described in Chapters 5 and 6.

Chapter 5

Generation of Robot Gesture

The realization of an ACE-based gesture generation framework for the humanoid
robot required implementations at two different levels (cf. Figure 4.12): first,
preparative adjustments were made with regard to the original ACE module, and
second, the complete ACE-to-robot interface connecting ACE with the WBM
controller had to be implemented. These accomplishments are described in the
following Sections 5.1 and 5.2 respectively. In light of the challenges faced when
generating conceptual motion as required by robot gesture, specific implementation
choices are highlighted (see RQ4). Finally, technical results obtained using the
implemented framework for robot gesture generation are presented and discussed
in Section 5.3.

5.1 Adjustments at the ACE Level

Prior to connecting ACE to the robot's WBM controller, a couple of preparative
modifications were made to the ACE module. These adjustments account for
the replacement of the original output platform in the form of the virtual agent
MAX with the targeted robotic platform. For this purpose, components of the
ACE system with inherent connections or dependencies to the body model of the
originally animated agent MAX were identified and tailored to conform to the
humanoid robot. Specifically, the skeleton description of the body model provided
for motion planning in ACE as well as the specification of gesture space had to be
redefined and adjusted as further described in Sections 5.1.1 and 5.1.2 respectively.
Figure 5.1 highlights the two adjusted components of the ACE module.

These modifications aimed at facilitating the implementation of the required
ACE-to-robot interface by pre-adjusting motion planning within the higher-level
module ACE. Alternatively, disregarding the mismatch between the different
embodiments at the ACE level would result in the complete allocation of the

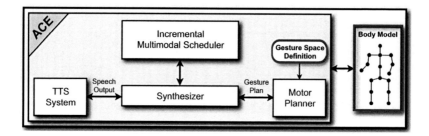

Figure 5.1: Adjusted components of the ACE module (framed in red): the skeleton description of the body model for motor planning in ACE (Section 5.1.1) as well as the specification of gesture space (Section 5.1.2) had to be redefined.

correspondence problem as illustrated in Section 4.3.2 to the lower-level interface. However, a realization at this level is more challenging. Therefore, adopting the chosen approach means that the motion retargeting problem can be encountered both at the preliminary planning level within ACE and, to a greater extent, at the execution level by the ACE-to-robot interface.

5.1.1 Skeleton Definition of ACE Body Model

As highlighted in Section 4.2.1 and illustrated in Figure 5.1, the ACE system is closely coupled with the kinematic body model of the embodied agent. Properties of the agent's kinematic skeleton are specified in a configuration file defining the limbs of the body as well as their associated body segments and joints. These, in turn, are further specified with regard to their respective types, rotations and translations, as well as their associated joint ranges and default joint angle values.

The skeleton definition file is specified in XML format and is read into the ACE system at start-up to build the corresponding body model for use at run-time. During program execution, the current joint configuration of the kinematic skeleton is constantly mapped onto the body model which is integrated into the visualization of the character animation. Conversely, the model provides ACE with a representation of the agent's body state and, based on differential quotients of the configurations of two successive animation frames, with a means to derive the joint velocities at any point in time. This proprioceptive feedback is used by the ACE motor planner for the state-dependent (de-)activation of respective motor control programs (MCPs) and local motor programs (LMPs) as described in Section 4.2.1 and illustrated in Figure 4.8.

In summary, the ACE body model plays an important role in motor planning and should thus depict the skeleton of the output platform as accurately as possible. However, in contrast to the animation of a virtual agent which is based on and directly linked to the state of the ACE body model, the current state of the robot cannot be as precisely reflected, since the robot is a physically independent platform. As such, the robot's state is further subject to other control primitives which are not accounted for by the ACE system, e.g., those instantiated by the collision avoidance as part of WBM control. Similarly, joint velocities derived from the ACE body model do not depict actual values of the physical robot. Nevertheless, adjusting the model to match the dimensions and characteristics of the robot's body is worthwhile, if only to conform with trajectory formation and motor planning so as to better fit with the robot's physical properties.

Accordingly, as part of the present work the XML file specifying the ACE body model was modified to encompass the robot's body segments and joints as well as their respective dimensions and motion ranges. Besides the general decrease in size and change of proportions, a number of joints originally modeled in the anthropomorphic body of the virtual agent were eliminated in the robot-specific model, since the latter comprises fewer DOF. The reduction in joints as well as the adjustment of segment dimensions for the definition of the robot's body model is visualized in **Figure 5.2**. Specifically, several joints including their connecting links in the spinal section of the virtual agent's skeleton were replaced with a single segment representing the robot's spine which comprises no additional DOF (v15 to vc7). Moreover, since the robot has no separate neck segment between its head and torso (cf. Figure 4.10), a single segment specifying the height of the head (vc4 to vc2) was defined as a substitute for the neck and head specified in the original MAX model.

Joint limits and DOF were further adjusted in the XML file by defining joint ranges which specify the lower and upper limit (in degrees) for each axis of every joint. For joints with fewer DOF in the robot model than in the virtual agent model lower and upper joint limits of the inflexible axes were each set to zero. For example, the wrist joint of the virtual agent comprises 3DOF, hence motion ranges were originally specified for the x-, y-, and z-axis respectively in the virtual agent's skeleton description. The robot, however, has only 1DOF in its wrist joint;

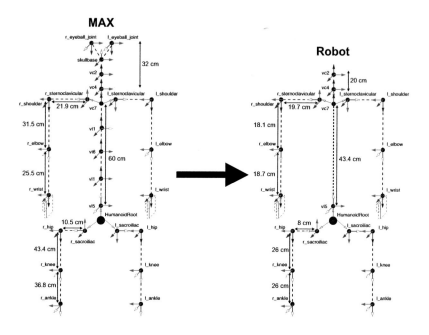

Figure 5.2: Reduction of joints and adjustment of segment lengths: the body model of the virtual agent MAX (left) was modified to account for the smaller number of DOF and kinematic dimensions provided by the robot (right).

this was accounted for by setting the lower and upper limits of the motion ranges for both the x- and y-axis to zero and by specifying a broader range of rotation about the z-axis.

The finger joints of the virtual agent's kinematic hand model (see Figure 4.7c) were not adjusted in the robot-specific model, as the robot only comprises 1DOF in the hand which inherently prohibits the control of single fingers. The robot's hand instead provides a grasp axis, allowing for the specification of an open or closed hand, or any intermediate configuration between the two. This limitation of the robot's hands, however, is difficult if not impossible to model in an adequate way within the XML definition of the ACE body model, since the definition can only specify general joint motion ranges but not additional interdependencies between the joints. As a result, attempting to provide an accurate hand model

106

at the ACE level would needlessly complicate matters while still not solving the correspondence problem at the execution level. Therefore, the mismatch of embodiment with regard to feasible finger movements is accounted for by the ACE-to-robot interface as described in Section 5.2.2.

5.1.2   Rescaling of Gesture Space Definition

The second aspect of adjusting the ACE module for use with a robotic output platform concerned the definition of robot-specific gesture space dimensions. As described in Section 4.2.1, the attributes of the "HandLocation" slot in a MURML specification (see Figure 4.5) define the relative location of the wrist based on the division of human gesture space as proposed by McNeill (1992, see Figure 4.6). This abstract representation format allows for the specification of the designated gesture stroke location based on the linguistic description of spatial features rather than concrete position vectors. The given combination of abstract location symbols as specified in a MURML file (e.g., LocShoulder LocRight LocFar) are then translated into numeric values by the ACE system at run-time.

In the original ACE framework used for the virtual agent MAX, the values assigned to the location symbols are based on gesture space dimensions of a human adult, which correspond to the characteristics of the agent's body model. Given the smaller scale of the humanoid robot, however, these values had to be adjusted to conform to the gesture space of the robot. For example, the longitudinal position symbol LocShoulder as specified for MAX defines a target at the shoulder height of the agent, i.e., at a height of 1.48 m. This value, however, would be located above the robot's head if directly applied and, in fact, could not even be reached by the robot's hand; it was therefore adjusted to correspond to the robot's shoulders at a target height of 0.95 m. Accordingly, all attributes and values of the "HandLocation" slot were rescaled and modified in the configuration file such that they specify the corresponding values in compliance with the robot's dimensions. These values were subject to testing and evaluation based on which further adjustments were made. This eventually led to the final configuration of a robot-specific gesture space definition.

5.2 Implementation of ACE-to-Robot Interface

Besides the adjustments made at the ACE level, the main work required for the proposed robot gesture generation framework involved the implementation of an interface that connects ACE to the humanoid robot via WBM control (cf. Figure 4.12). To combine these two systems, several aspects had to be taken into consideration, which are highlighted and elaborated in the following sections.

5.2.1 Control Strategy for Trajectory Formation

The present work on robot gesture control using the ACE system offers two different control strategies for driving the robot using motor commands derived from the virtual agent framework: by establishing correspondence between the ACE body model and the body of the physical robot either in joint space or in task space. The first method involves an extraction of the joint angles from the kinematic body model of ACE which are then mapped onto the robot body model controlled by the WBM system. The second method amounts to using ACE to formulate a trajectory in terms of end-effector targets in task space, based on which a joint space description can be derived by the inverse kinematics (IK) module of the robot's WBM controller.

For several reasons elaborated in the following, the choice was made in favor of the second approach, namely **task space control**. Since ACE was originally designed for a virtual agent application, it does not entirely account for certain physical restrictions such as collision avoidance which, however, are crucial in the control of a physical robot. For example, while it is possible and acceptable for the virtual agent's arms to rest against its body, any form of self-touch is not admissible for the robot. Consequently, joint angle configurations – especially those of the shoulder – as specified for the virtual body model in ACE frequently lead to joint states that are not feasible on the robot, since they would result in self-collisions. In fact, whenever such infeasible joint angle commands are sent to the robot, the real-time collision avoidance algorithm integrated into the WBM software overwrites the given target configuration with an optimized solution. This motivates the necessity of adopting the task space approach, since such automatic modification of joint angles, as implied by the first method, would be difficult to control and regulate externally by the ACE-to-robot interface; in the worst case, this may result in a deviation of the desired end-effector trajectory. In contrast, when employing the second method, i.e., controlling the robot via end-point trajectories based on which IK is solved using the robot's WBM controller, feasible and thus safer robot postures are ensured.

The above mentioned drawbacks with regard to a joint space approach lead to the following question: which features of gestural movement are essential in conveying the correct meaning of a gesture? And more specifically in this context: what information has higher priority - precise joint configurations or accurate end-effector positions? Findings from human perception studies suggest that humans largely track the hand or end-points of one another's movement, even if the movement is performed with the entire arm (Mataric and Pomplun, 1998). This was found to be true even in observational learning scenarios, i.e., when the task was not only to observe but also to imitate the perceived motor action. These and other findings supporting the *goal-directed theory of imitation* (Wohlschläger et al., 2003) suggest that observing end-point information is sufficient to successfully imitate goal-directed movements (Maslovat et al., 2010). In effect, during movement observation, such prioritization of task level information is in line with empirical evidence showing that humans generally plan goal-directed motion in the lower-dimensional task space rather than in joint space.

Given these biologically as well as empirically motivated arguments, the reasons for choosing task space over joint space control can be summarized as follows.

- From an intuitive, biologically inspired gesture *generation* point of view, trajectory formation in task space complies with the human approach to high-level planning of goal-directed movements (e.g., gestures) in external coordinates.

- From a gesture *perception* point of view, empirical evidence as discussed above suggests that even with a deviation of joint angles the form and meaning of a gesture can typically still be conveyed. To humans observing goal-directed movements such as gesture, the end-points of the arms are more important than the individual joints. Those rare cases in which a communicative gesture may demand an exact joint configuration do not justify the complexity involved in realizing such a system. In addition, given the robotic platform used, such an approach does not guarantee completely natural and lifelike gesture appearance.

- From a technical *realization* point of view, choosing task over joint space control facilitates the implementation of the designated gesture generation framework, since it enables a straightforward solving of the correspondence problem. Furthermore, joint space control may result in the robot's self-collision avoidance mechanism by autonomously altering the joint angle value originally assigned; task space control mitigates this issue by instead prioritizing the accuracy of the end-effector position. The WBM controller can then be used to solve IK autonomously while finding the most suitable and feasible collision-free joint configuration for the robot.

Consequently, the proposed ACE-to-robot interface was implemented to extract the **wrist position and respective hand orientation** of the virtual agent's gesturing arm in external coordinates from the ACE body model. These coordinates are then transformed from the body-centered external space of the ACE body model to that of the robot. Despite the adjustments made at the ACE level (see Section 5.1), this transformation is necessary, as the two platforms specify different origins for their world coordinate systems. In ACE, world coordinates originate in the hip of the agent's body (see `HumanoidRoot` joint in Figure 5.2), whereas in the kinematic body model of the robot's WBM controller, the origin is defined in the heel coordinate system (see Figure 4.10). During the transformation, an appropriate offset along the z-axis is therefore used to account for the height of the robot's legs. In contrast, the hand orientation of the virtual model is directly mapped onto the wrist angle of the robot.

5.2.2 Mapping of Hand Shapes

With only 1DOF in each hand, the humanoid robot is more limited in performing single finger movements than the virtual character originally animated by ACE. As illustrated in Figure 4.10 and described in Section 5.1.1, the DOF in the robot's hand provides a grasp axis which allows for the specification of a closed or open hand as well as any intermediate configuration between the two. To this end, the designated hand configuration is defined by means of a single joint value, for example, specifying a grasp angle of zero results in an open hand.

Such limited expressiveness with regard to the robot's hands makes it difficult to model and realize a wide range of different hand gestures and finger postures. This limitation was countered by specifying three basic hand shapes as depicted in **Figure 5.3** which can all be performed by the robot: open hand, pointing with the index finger, and closed hand. These finger constellations are assigned to the three basic symbols available to the `"HandShape"` slot in a MURML specification

Figure 5.3: Different hand shapes specified for hand gesture generation on the humanoid robot: *open hand, pointing with the index finger,* and *closed hand* (reprinted from Salem et al., 2010c).

(cf. Figure 4.5), namely `BSflat`, `BSffinger`, and `BSfist` respectively.

Any hand gesture derived from the ACE body model that deviates from these three basic symbols, e.g., by defining specific finger constellations not feasible on the robot, are realized on the robot by using an open hand shape. This way, MURML files originally specified for the virtual agent can still be interpreted and executed when utilized in the transferred framework. Ideally, however, when specifying MURML descriptions for behavior generation in the robot-specific gesture generation framework, only the three above hand shape symbols should be explicitly used, as these can be directly mapped onto the robot.

5.2.3 Sampling Rate

The ACE-to-robot interface connects behavior planning and scheduling in ACE with the WBM controller of the robot (cf. Figure 4.12). For this purpose, the interface extracts task space commands from ACE (as specified in Section 5.2.1), transforms them into the robot's world coordinate system, and issues them to the WBM controller which sets the robot into motion. In addition to these efferent control signals, afferent feedback is integrated into the control architecture to monitor possible deviations of actual robot motor states from the kinematic body model provided by ACE.

The required bi-directional interface is realized by a feedback-based closed loop in which motor commands are transmitted on the condition that at least one local motor program (LMP) for gesture generation is active in ACE. The interface updates the kinematic body model coupled to ACE as well as the internal model of the robot in the WBM controller at a sample rate r. This process synchronizes two competing sample rates to ensure successful interoperability: firstly, that of the ACE system, and secondly, that of the WBM software controlling the robot. For this purpose, any of the following mapping rates can be employed:

1) **Sampling at target positions.** ACE sends only the final positions and orientations of movement segments and delegates the trajectory formation and respective movement generation entirely to the robot's WBM controller.

2) **Sampling at each n-th frame.** ACE sends control parameters regarding the current wrist position and hand orientation of the agent's body at a fixed rate to the robot's WBM controller.

3) **Adaptive sampling rate.** ACE tethers the WBM controller using different sampling rates, ranging from one sample per frame to taking only the end positions, depending on the complexity of the trajectory.

To illustrate, if the trajectory is linear, strategy 1 above can be expected to serve as the best mechanism, since only distance information would likely be required. If, on the other hand, the trajectory is complex, strategy 2 can be expected to be optimal, since a sequence of small movement vectors would likely be required to guide the robot controller. If, however, the gesture is formed from different types of sub-movement, e.g., a linear trajectory for gesture preparation and a curved trajectory for the stroke, the combined approach of strategy 3 using an adaptive sampling rate can be considered the most advantageous.

With regard to the present implementation, the first method was discarded as it fails to adequately account for non-linear (e.g., curved) trajectories. For example, if a circular gesture is to be performed, it can be defined by four key-points to be connected by four curved sub-trajectories. However, sending these key-points to the WBM controller using the first mapping strategy would result in a square rather than round gesture trajectory.

For successful handling of such cases, in the realized framework the second method was implemented with a maximal sampling rate in which each successive frame of the movement trajectory is sampled and transmitted to the robot controller ($n = 1$). Given a frame rate of 20 frames per second, which is flexibly adjustable with ACE and compliant with the update rate of 5 ms on the part of the robot controller, this results in a large number of sample points. These, in turn, ensure that the robot closely follows the potentially complex trajectory planned by ACE. Based on the technical results presented in this thesis, this method was shown to be a viable approach which meets the technical requirements of the present work. Given the real-time planning and generation capacities of ACE as well as its potential to produce complex gesture trajectories, such tight coupling with the robot controller was considered a favorable solution, as it allows for maximal controllability of the robot.

Alternatively, using the third strategy would allow for adjusting the sampling rate depending on the trajectory's complexity, which may vary from simple straight movements, e.g., for gesture preparation or retraction, to complex curved shapes for the gesture stroke phase. Whether or not this strategy would lead to improved results for the generation of robot gesture in combination with ACE remains a point of future investigation.

5.2.4   Outline of the Control Architecture

Figure 5.4 illustrates the control architecture for robot gesture generation based on the previously described implementation choices. The hierarchically organized framework combines conceptual behavior representation and planning in ACE with motor control primitives for hand and arm movements of the physical robot body.

As depicted in the figure, the generation process of the efferent control pipeline is initiated by a MURML specification defining a timed gesture plan. Based on the given movement constraints and the current state of the ACE body model, the motor planner in ACE determines a suitable movement trajectory. Once a LMP has been activated, the transfer of motor commands to the ACE-to-robot interface is triggered. At each time step t, which corresponds to the sample rate r, the interface receives the current wrist position as well as the hand shape and orientation of the ACE body model ($\vec{x}_{ace}$). The interface is coupled to the kinematic body model of the robot via a simulation environment of the WBM controller which reflects the current state of the robot. The ACE-specific vector $\vec{x}_{ace}$ is transformed into the robot's world coordinate system, while the specified hand configuration from ACE is mapped onto one of the three robot hand shapes depicted in Figure 5.3, together yielding the robot-specific vector $\vec{x}_{rob}$. This task space target is forwarded to the WBM controller which solves the IK problem of the arm on the velocity level. The resulting joint space description $\vec{\theta}_{rob}$ of the designated trajectory point is then applied to the real robot for execution.

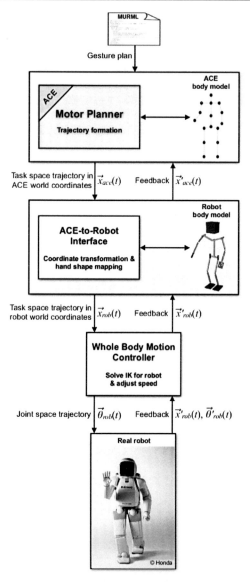

Figure 5.4: Control architecture for the realization of robot gesture: the framework combines conceptual behavior representation and planning in ACE with motor control primitives for hand and arm movements of the physical robot body.

The afferent feedback pipeline emanating from the robot transmits the actual current wrist position and complete joint configuration of the robot's arm to the WBM controller $(\vec{x}'_{rob}, \vec{\theta}'_{rob})$. The robot-specific vector $\vec{x}'_{rob}$ is then passed on to the ACE-to-robot interface which transforms the actual wrist position and hand orientation of the real robot into the ACE world coordinate system. The resulting vector $\vec{x}'_{ace}$ is finally sent back to the ACE module, in order to monitor potential deviations of the actual position and orientation of the robot's hand from the motor states of the kinematic body model associated with ACE.

The realized bi-directional behavior generation pipeline couples the ACE framework with the perceptuo-motor system of the humanoid robot, thus enabling the robot to flexibly produce communicative gestures at run-time. Technical results presented and discussed in the following section were produced using the outlined robot gesture generation framework.

5.3 Technical Results and Discussion

To evaluate the general performance of the implemented gesture generation framework and the appearance of the produced robot gestures, preliminary results were produced in a feedforward manner. That is, although sensory feedback was available to the system, it was not used to modify the generated movement behavior at run-time. This way, the original functioning as well as potential limitations of the framework could be best identified before envisaging gesture synchronization with the more constraining modality of speech.

To obtain the results presented in this section, commands indicating the wrist position as well as the hand orientation and hand shape of the ACE body model were transmitted to the robot in real-time at a sample rate of 20 frames per second. A wide range of gesture types, one-handed as well as two-handed, with both static and dynamic gesture strokes were generated based on dedicated MURML specifications. In the following, two representative examples are illustrated and discussed in more detail, providing an insight into the overall performance of the system. Specifically, in Section 5.3.1 technical results derived from a feature-based MURML description of a one-handed gesture with a static gesture stroke are discussed. Additionally, an example of a two-handed gesture with a dynamic gesture stroke is presented in Section 5.3.2.

5.3.1 One-handed Gesture with Static Stroke

As outlined in Section 2.1.3, a gesture can either have a dynamic or a static gesture stroke which expresses the meaning of the gesture. In the case of the first, the hand moves dynamically from a designated stroke start position to a different target position. In contrast, in the case of a static stroke, the gesturing hand remains motionless at the gesture target position, typically for the duration of the affiliate. Such stroke hold, for example, is often observed in deictic gestures.

To evaluate the performance of the robot when generating a one-handed static gesture, the MURML specification depicted in **Figure 5.5** was used as input for the robot control architecture outlined in Section 5.2.4. The resulting gesture output is presented in **Figure 5.6** in which the robot is shown next to a panel displaying the current state of the internal WBM robot body model and the ACE kinematic body model respectively at each time step. The screenshot sequence reveals that the physical robot is able to perform the designated gesture fairly accurately but with some inertial delay compared to the internal ACE model.

This observation is supported by **Figure 5.7a** in which each dimension of the wrist trajectories of the ACE body model and the robot are plotted against time. The depicted delay results from the step response during motion acceleration as well as the more restrictive velocity limits imposed on the physical robot platform. Specifically, the plot of the z-axis of the robot's trajectory (blue dotted line), which represents the vertical rising of the arm and thus the most prominent dimension of this gesture, indicates that the robot needs ~2.3 seconds to reach the stroke target. In contrast, the plotted trajectory of the ACE body model (red solid line) reveals a much faster acceleration toward the stroke position, reaching the target

```
<definition><utterance>
 <behaviorspec id="example_static">
  <gesture>
   <constraints>
    <parallel>
     <static slot="HandShape" value=" BSflat"/>
     <static slot="ExtFingerOrientation" value="DirA"/>
     <static slot="PalmOrientation" value="DirL"/>
     <static slot="HandLocation" value="LocShoulder LocPeripheryRight LocStretched"/>
    </parallel>
   </constraints>
  </gesture>
 </behaviorspec>
</utterance></definition>
```

Figure 5.5: MURML specification used to generate the one-handed static gesture depicted in Figure 5.6.

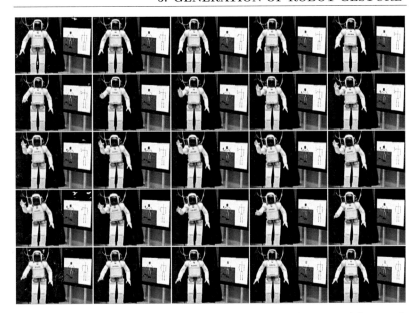

Figure 5.6: One-handed static robot gesture realized with the proposed framework; for comparison, the physical robot, WBM robot body model, and the kinematic ACE body model are shown (left to right, top-down, sampled every four frames (0.16 sec)).

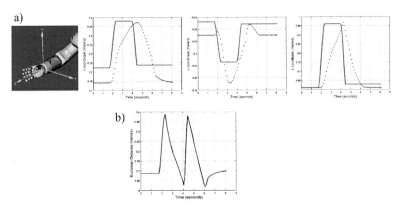

Figure 5.7: a) Plots of x-, y-, and z-axis respectively (in robot world coordinates) of the right wrist positions of the ACE body model (*red, solid*) and the robot (*blue, dotted*) during gesture execution; b) Euclidean distance between the current wrist positions of the ACE body model and the robot over time.

117

within just ~0.5 seconds, while the robot's wrist is still moving toward it. The last two pictures of the first screenshot row displayed in Figure 5.6 visualize this discrepancy. Consequently, the wrist position of the robot only 'catches up' with the ACE body model at the end of the stroke hold when, in fact, the retraction phase has already begun, resulting in a shorter stroke phase for the robot.

Figure 5.7b further emphasizes this observation with a graph of the Euclidean distance between the current wrist position of the robot and its corresponding target position as specified by the ACE body model plotted against gesture execution time. It indicates that the greatest discrepancies between the actual and the target position occur in the gesture preparation and retraction phases during which the ACE model exhibits greater movement speed.

Despite the general limitation in speed, these findings substantiate the feasibility of the proposed approach. The presented gesture example shows that the implemented method of task space control of the robot, i.e., disregarding the joint angles as generated in ACE, does not impair the overall shape and meaning of the gesture. Moreover, the output example of the one-handed gesture as depicted in Figure 5.6 demonstrates the effect of the balance controller that is integrated into the WBM system: the one-sided motion of the upper body toward the gesture target position leads to a displacement of the overall center of gravity. This displacement is compensated by the balance controller, resulting in an upper body shift and a compensating movement of the legs which can be observed particularly in the third and fourth row of the screenshot sequence. Most importantly, however, the images show that the balancing movement of the lower part of the robot's body does not significantly affect the appearance of the gesturing robot. Since a human observer typically focuses on the end-point of the effector (cf. Section 5.2.1; Mataric and Pomplun, 1998), these secondary movements of the legs are not expected to distort the communicative intent of the gesture.

5.3.2 Two-handed Gesture with Dynamic Stroke

In contrast to gestures with a static stroke, the meaning-bearing phase of dynamic gestures consists of a stroke movement which begins at one hand location and terminates at another. In order to evaluate the performance of the robot when generating a two-handed gesture with a dynamic stroke, the MURML description illustrated in **Figure 5.8** was used as input for the implemented robot control architecture. It specifies a gesture in which both hands are simultaneously moved

```
<definition><utterance>
 <behaviorspec id="example_dynamic">
  <gesture>
   <constraints>
    <symmetrical dominant="left_arm" symmetry="SymMS">
     <parallel>
       <static slot="HandShape" value=" BSflat"/>
       <static slot="ExtFingerOrientation" value="DirA"/>
       <static slot="PalmOrientation" value="DirPD"/>
       <dynamic slot="HandLocation">
         <dynamicElement type="linear">
         <value type="start" name="LocShoulder LocPeripheryLeft LocNorm"/>
         <value type="end" name="LocAbdomen LocPeripheryLeft LocNorm"/>
         </dynamicElement>
       </dynamic>
     </parallel>
    </symmetrical>
   </constraints>
  </gesture>
 </behaviorspec>
</utterance></definition>
```

Figure 5.8: MURML specification used to generate the two-handed dynamic gesture depicted in Figure 5.9.

downwards from shoulder height to the level of the abdomen during the dynamic stroke phase.

The resulting gesture output is presented in **Figure 5.9**. Again, the screenshot sequence reveals that the performance of the robot lags behind that of the internal ACE body model. In particular, the last picture of the first row illustrates how the ACE model reaches the start position of the dynamic gesture stroke, while the robot is still in the early preparation phase. In this case, however, due to the dynamic properties of the gesture stroke, the inertial delay of the robot not only results in a shortened stroke phase, but notably, it also alters the path of the dynamic stroke trajectory. As the target position transmitted to the robot is updated every 500 ms, the stroke start position has already been overwritten by the time the robot has completed its gesture preparation phase. This means, once the robot is ready for the gesture stroke onset, the updated target position, as issued to the robot based on the current position of the ACE body model, is already that of the final stroke position. As a result, the gesture stroke performed by the robot does not begin at shoulder height as specified by the MURML description, but slightly above the abdominal area.

This process is visualized in **Figure 5.10a**. With particular reference to the plot of the z-coordinate, which depicts the vertical movement of the wrist of the robot's right arm, the plotted trajectory of the ACE body model (red solid line) distinctly marks the stroke onset at the ~2.5 seconds time mark, followed by the stroke movement (~2.5 sec to ~3.1 sec) and a subsequent post-stroke hold (~3.1

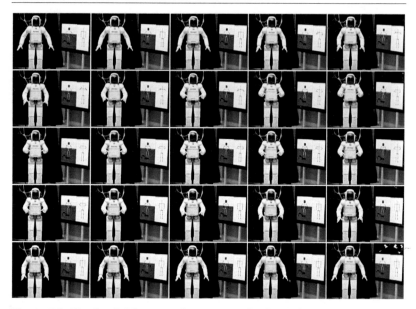

Figure 5.9: Two-handed dynamic robot gesture realized with the proposed framework; for comparison, the physical robot, WBM robot body model, and the kinematic ACE body model are shown (left to right, top-down, sampled every four frames (0.16 sec)).

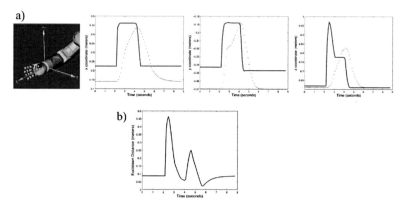

Figure 5.10: a) Plots of x-, y-, and z-axis respectively (in robot world coordinates) of the right wrist positions of the ACE body model (*red, solid*) and the robot (*blue, dotted*) during gesture execution; b) Euclidean distance between the current wrist positions of the ACE body model and the robot over time.

sec to ∼4.0 sec). In contrast, the robot's trajectory (blue dotted line) bears a simple bell-shape which is characterized by an overshoot around the time the robot 'catches up' with the ACE body model at the post-stroke hold phase. This overshoot is likely caused by the sudden shift in movement direction as transmitted by the ACE body model during execution of the dynamic gesture stroke.

Figure 5.10b depicts this discrepancy in the two trajectories by plotting the Euclidean distance between the robot's actual wrist position and the target position as dictated by the ACE model over time. As with the one-handed gesture (Section 5.3.1), the greatest mismatch between the robot's actual and target position occurs in the gesture preparation and retraction phase respectively. As seen in the plotted ACE trajectory, these phases are characterized by significant acceleration of movement speed at motion onset and abrupt shifts in direction. Problematically, in the present example of a dynamic gesture, this leads to a distortion of the gesture stroke, since the first part of the dynamic movement is missed by the robot. The main source of the problem lies in the extreme acceleration capacity of the ACE model at gesture preparation and retraction; this is exemplified by all ACE graphs in Figure 5.10a in which steeply rising and sloping curve progressions are evidenced.

Generally, in ACE the arm movement toward the stroke onset position in the preparation phase as well as the movement from the stoke end position back to the default position during the retraction phase are generated as linear trajectories. As such, their movement duration is estimated based on *Fitts' Law* (Fitts, 1954), a psychological model of human motor behavior. It describes the time required for a rapid goal-directed movement to a target location as a function of the distance to the target and the size of the target. Fitts' Law has been formulated mathematically in a number of different ways; in the ACE system, it was implemented according to the Shannon formulation proposed by MacKenzie (1992), which defines the predicted *movement time (MT)* as follows:

$$MT = a + b \, log_2 \left(\frac{A}{W} + 1 \right) \tag{5.1}$$

whereby coefficient a denotes the intercept and b the slope (both to be empirically determined), A represents the amplitude (i.e., the distance to the target), and W the width of the target. In the original ACE implementation, the intercept coefficient a is set to zero, the slope coefficient b has the value 0.12, while the target width W is set to 1 (i.e., W is neglected).

Since Fitts' Law is based on the study of human movement trajectories, it may be suitable for the modeling of motor behaviors of computer animated virtual agents which are not physically restricted in movement speed. But in view of the present results, when applied to a robotic platform, the predefined values as specified in ACE do not appear to comply with the existing physical constraints. To account for the discrepancy in movement speed with regard to the robotic output platform, the time estimation applied to plan and generate gesture trajectories in ACE requires adjustment for use with the robot.

In an initial attempt to address this issue, the time estimation function in ACE was experimentally fitted to decrease the speed of the generated target trajectories. **Figure 5.11** illustrates a selection of results based on the execution of the dynamic gesture depicted in Figure 5.9 with altered gesture speed; corresponding to the third graph in Figure 5.10a, the trajectory along the z-axis is plotted against time. Specifically, Fitts' Law (Equation 5.1) was adjusted by setting the slope coefficient b to values 18, 20, and 22 respectively (Figure 5.11a). In an alternative approach, Fitts' Law was replaced with a simple time estimation function specifying the average velocity for gesture execution on the robot as 350 mm/sec, 300 mm/sec, and 250 mm/sec respectively (Figure 5.11b). In comparison to the observations

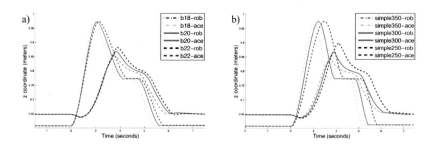

Figure 5.11: Experimental fitting of estimated gesture execution time: plots of the right wrist trajectory along the z-axis (cf. Fig. 5.10a) of the ACE body model (red shades, 'ace' label) and of the robot (blue shades, 'rob' label) by a) adjusting **Fitts' Law** coefficient b with values 18, 20, and 22 respectively; b) using a **simple time estimation** based on average velocity values 350 mm/sec, 300 mm/sec, and 250 mm/sec respectively.

made in Figure 5.10a, the plots indicate that decreasing gesture speed in ACE results in the robot's trajectory being more closely aligned to the target trajectory. Crucially, although still disclosing deviations, the gesture phases and the dynamic stroke are better preserved than with the initial configuration. Interestingly, with regard to the plotted example and as far as it is visible to the naked eye, the simple time estimation approach appears to be just as suitable if not better than the adjusted Fitts' Law approach.

On a more general note, it should be kept in mind that ultimately, the goal of the present work is not to implement a robotic 'clone' of the virtual agent MAX by aiming at almost identical behavior. Rather, the objective is to employ ACE as a sophisticated tool to generate multimodal robot behavior for a humanoid robot. That is, the aim is to customize the virtual agent framework into a re-engineered software platform for co-verbal robot gesture generation, potentially by altering central ACE functionalities. Thus, even if the robot's gesture trajectory does not exactly match the ACE trajectory, it matters little, since the crucial demand is to convey the intended meaning. As the behaviors generated by the robot will not be ultimately benchmarked in comparison to the ACE model, it is crucial that a given robot gesture is perceived as meaningful in 'stand-alone' execution.

Moreover, given the ultimate technical goal of the present work, namely the coordination of the generated gestures with concurrent speech, such adjustments also greatly concern the aspired synchronization of the two modalities. Therefore, this aspect of movement planning and timing should not be analyzed in isolation and is thus addressed in more detail in the context of speech-gesture synchronization in Chapter 6.

The results presented in this section demonstrate the potential as well as the limitations of the implemented solution. The resulting framework allows for the generation of robot gesture from arbitrary MURML-based gesture specifications. Importantly, they show that neglecting joint angle information as generated in ACE does not impair the overall shape of the gesture generated by the robot. Thus, controlling the robot via task space commands provides an adequate and safe way to generate arm movements for the robot. While the outlined limitations require careful consideration especially in view of gesture synchronization with speech, the overall appearance of the gesturing robot reveals promising potential for lifelike gestural behaviors.

5.4 Summary

The focus of this chapter was on the realization of a control architecture for the generation of robot gesture based on the ACE system. The implementation of the designated framework was shown to comprise two conceptual components: first, adjustments made at the ACE level, and second, the realization of an interface connecting ACE with the WBM controller of the robot.

Initially, in **Section 5.1** modications of the ACE module and the corresponding body model were described. These changes aimed to account for the replacement of the original output platform, the virtual agent MAX, with the targeted robotic platform. For this purpose, the skeleton definition originally specified for the virtual agent was modified to match the kinematic dimensions and properties of the humanoid robot (Section 5.1.1). In addition, the gesture space definition in ACE, which was originally specified with regard to the skeleton of the agent MAX, was rescaled and adjusted to match the robot's dimensions (Section 5.1.2).

Section 5.2 provided detailed information on the implementation of the ACE-to-robot interface, the core component of the technical framework for robot gesture generation. In this context, a number of implementation alternatives and decisions made in the realization process were outlined and discussed (see RQ4). In particular consideration of various aspects of the correspondence problem, task space control was identified as an appropriate control strategy for trajectory formation of the robot's gestures (Section 5.2.1). Consequently, it was explained how the ACE-to-robot interface was implemented to extract wrist positions and respective hand orientations in external coordinates from the ACE body model to then map them onto the robot. In addition, given the smaller number of DOF in the robot's hands compared to the agent's hands, specifications of hand and finger configurations were introduced, detailing how they were modeled to match one of three basic hand shapes feasible on the robot (Section 5.2.2). These extracted movement features were highlighted as being subject to an adequate transfer and mapping rate between the virtual agent framework and the robotic platform (Section 5.2.3). A constant sampling rate was indicated as the chosen method for implementation, meaning that control parameters and feedback regarding the current wrist position, hand orientation, and hand shape of both the agent and the robot are updated at every frame. Based on these implementation choices, the resulting hierarchical control architecture for robot gesture generation based on ACE was illustrated and explained (Section 5.2.4).

Finally, in **Section 5.3** preliminary technical results produced using the implemented framework were presented based on two representative examples. Specifically, results derived from MURML descriptions of a one-handed gesture with a static gesture stroke (Section 5.3.1) as well as of a two-handed gesture with a dynamic gesture stroke (Section 5.3.2) were depicted and discussed. The outlined results demonstrated not only the capacities but also the limitations of the system, thus highlighting potential challenges with regard to successful synchronization of the generated gestures with speech. In this way, the implementation work and results described in this chapter provide a conceptual basis and grounding for the technical aspects addressed in the following chapter.

Chapter 6

Synchronization of Robot Gesture with Speech

While the previous chapter focused on the generation of robot gestures alone, this chapter centers on the synchronization of produced gestures with speech. Given the limitations identified at the gesture generation level, movement synchronization with another constraining modality is particularly challenging: in addition to form feature constraints inherent in the gesture itself, timing constraints imposed by accompanying speech further influence the execution of the desired movement.

In view of these specific requirements and further challenges highlighted in Section 4.3.3, the concept of a multimodal scheduler that ensures fine synchronization of robot gesture with speech is introduced in Section 6.1 of this chapter. The presented scheduler extends the original ACE component with two essential functionalities. Firstly, as further described in Section 6.2, it provides a predictive forward model which has been tailored to the requirements of the robotic platform at hand. Secondly, as specified in Section 6.3, it incorporates a feedback-based adaptation mechanism which allows for an on-line adjustment of the synchronization of the two modalities at run-time. Finally, results of the two extended features are presented and discussed in the respective sections.

6.1   Concept of Extended Multimodal Scheduler

Findings from human gesture research reveal that the onset of the co-verbal gesture stroke generally precedes or, at the latest, begins at the onset of the nucleus of the conceptual affiliate in speech (cf. Section 2.2.1). Ensuring this empirically validated requirement for co-expressive synchrony between the two modalities poses a major challenge for artificial communicators such as virtual agents or social robots. As reviewed in Chapter 3, up to the present time most technical systems dedicated to co-verbal gesture generation achieve synchronization by means of gesture adaptation to the structure and timing of running speech.

While this may represent an acceptable solution to realize multimodal behavior for virtual agents, such unidirectional adaptation may yield mistimed synchronization on a physically more constrained robotic platform. Although the ACE system is more sophisticated than other virtual agent frameworks in that it also allows for speech adaptation to gesture timing between two successive chunks (see Section 3.1.1), it is still similarly restricted at the intra-chunk level. Moreover, both gesture and speech generation are performed ballistically in a feedforward manner, that is, once the behaviors have been planned and scheduled, they cannot be re-adjusted during execution. This is particularly true of all existing robotic applications proposing technical models for speech-gesture generation, as they currently lack the potential for closed-loop control (see Section 3.2.1).

In an attempt to overcome some of the issues identified in other systems – but also in the ACE framework – the scheduler proposed in this chapter represents an extended and improved version of the ACE scheduler originally developed for application with virtual agents. Details on the conceptual realization and on implementation decisions made are provided in the following.

6.1.1 Speech Synthesis

For the generation of speech output, the open source text-to-speech synthesis system *MARY* (Modular Architecture for Research on speech sYnthesis; Schröder and Trouvain, 2003) Version 3.6.0 was used. It features a modular design and an XML-based internal data representation and is currently utilized as the standard text-to-speech (TTS) system for the ACE framework. Several languages including English and German are supported, thus providing the option to flexibly switch between different languages for speech output.

Furthermore, the TTS system supports various speech synthesis technologies including *unit selection* and *HMM-based synthesis* (Pammi et al., 2010). Unit selection builds on a large speech database out of which units of variable size which best correspond to the target utterance are selected (Schröder, 2004). Alternatively, for HMM-based synthesis, Hidden Markov Models are trained a priori with a data set in order to derive the characteristics of a specific voice (Tokuda et al., 2004). A major advantage of this method over concatenative approaches lies in its flexibility, since HMM-based synthesis does not require a speech database and further allows for the modification of voice characteristics by adjusting a number of HMM parameters. For the realization of both German and English speech output within the scope of the present work, predefined HMM-based voices from the MARY software package were employed.

Generally, four processing steps of the TTS system can be distinguished[1]: preprocessing, natural language processing (NLP), calculation of acoustic parameters, and synthesis. Most relevant to the present work are the last two of these steps, namely the calculation of parameters, during which every phoneme is assigned a duration in milliseconds, as well as the actual synthesis of the speech output, which transforms the derived acoustic parameters into an audio file.

Speech synthesis and multimodal scheduling of the ACE framework are tightly coupled with these two final processing stages of the MARY system. Once the <specification> tag defining the spoken utterance of a given MURML file has gone through the preprocessing and NLP stages, the TTS system provides ACE with a complete list of phonemes and their respective durations before generating the speech output file at the final processing stage. The phoneme duration values determined by the MARY system provide the ACE scheduler with speech-related timing information based on which the complete multimodal utterance is planned and the accompanying gesture is scheduled. The audio file generated by the TTS module is finally replayed by command of the ACE scheduler at run-time, ideally in temporal synchrony with the accompanying gesture. The underlying synchronization mechanism is further described in the following sections. For more details on the MARY text-to-speech system see Schröder and Trouvain (2003), Schröder (2004), and Pammi et al. (2010).

6.1.2 Limitations of Original ACE Scheduler

As part of the present work, the first step toward multimodal behavior generation for a robot based on ACE consisted of the use of the original scheduler provided by the framework to assess its initial performance and then decide whether it would require further modification. Given the examples of technical results of generated robot gestures presented in Section 5.3 of the previous chapter, the predefined movement speed of target trajectories was already identified as requiring

---

[1]http://mary.dfki.de/documentation/overview/ – accessed March 2012

adjustment. Therefore, initially the function estimating the gesture execution time in ACE was experimentally fitted in a similar fashion as described at the end of Section 5.3.2 and as illustrated in Figure 5.11, however, this time while generating accompanying speech output.

Although providing closer approximations to the original ACE trajectory, slowing down the robot's movement speed too much was found to result in 'lethargic'-looking gesture behavior. Thus, a compromise between accurate trajectory replication and lively appearing gesture behavior was to be found which, in addition, would synchronize well with accompanying speech. Exploratory tests with a large variety of multimodal MURML utterances yielded an acceptable approximation of co-verbal gesture execution time with slope coefficient b of the Fitts' Law equation (5.1) set to the value 18.[2] Alternatively, a simple time estimation assuming 300 mm/sec as average motion velocity for robot gesture generation was identified as another viable setting for initial use.

Despite yielding acceptable temporal synchrony between robot gesture and speech in a wide range of tested utterances, tests using distant gesture targets or sentences that start with the affiliate revealed that synchronization was not always optimal. That is, in some cases the gesture stroke was only performed *after* the affiliate. Therefore, a more flexible multimodal utterance scheduler was required in order to allow for a finer cross-modal adaptation between robot gesture and speech. Essentially, the ACE scheduler originally developed for application with virtual agents was identified as lacking the two following major functionalities, particularly when used on a robotic platform:

1) **Accurate prediction function for movement timing**. Being designed for an animated virtual character, the predictive model implemented in ACE employs Fitts' Law as a basic estimation function to anticipate the time needed for the agent to perform a body movement. In this regard, the value used for the slope coefficient b has not been empirically determined, since it was sufficient in the virtual agent environment to set an arbitrary value based on which the animation can be generated. However, given real physical constraints and limited joint velocities, such approximation to estimate gesture motion time has proven inadequate for action generation with a humanoid robot.

---

[2]Although this value was first merely assessed with the naked eye, it was subsequently further evaluated within the scope of the experimental studies described in Part III of this thesis.

2) **Feedback-based, cross-modal adaptation mechanisms at intra-chunk level.** In the original ACE scheduler, within a chunk of multimodal behavior production, gesture execution is scheduled to match the timing of speech. Once the multimodal behaviors have been planned and scheduled, no further adjustments can be made based on feedback during execution. However, this lack of cross-modal adaptability within a chunk and the ballistic generation of complete gesture and intonation phrases conflict with psychological findings from gesture research that suggests such cross-modal interactions (e.g., de Ruiter, 1998; Kendon, 2004). While nevertheless a feasible approach in a virtual agent application which does not typically face any unforeseen physical limitations or critical deviations in the animation of body movements in predetermined speed, it proves problematic when applied to a robot. Given the physical platform, the robot's gesture execution times are difficult to predict accurately and may deviate from the scheduled time during execution. Therefore, to prevent mistimed synchronization, it is not sufficient to adapt only one modality (i.e., gesture) to the other (i.e., speech) within a chunk, especially if execution of arm and hand gestures cannot be performed with arbitrary speed.

The technical requirements and extended features of an improved multimodal scheduler arising from the above described limitations of the original ACE scheduler are elucidated in the following sections.

6.1.3 Extended Features

As visualized in Figure 3.7, the incremental multimodal scheduler provided by the original ACE framework augments the classical two-phase 'planning-execution' procedure with additional phases of the speech-gesture production process. For the sake of simplicity, however, the traditional breakdown into planning and execution phase is used to illustrate the extended features of the improved scheduler proposed in this chapter. Such simplified distinction is further supported by the fact that each of the two major limitations of the old scheduler listed in Section 6.1.2 refers to one of these two phases.

In view of the above described limitations, an improved multimodal scheduler needs to address the following two questions previously highlighted as expected challenges at the speech-gesture synchronization level (see Section 4.3.3). First, with regard to the **planning** phase, how can a more reliable prediction of robot-specific gesture execution time be obtained to allow for better multimodal scheduling? Second, with regard to the **execution** phase, how can planning and scheduling errors be accounted for during execution if they were to cause mistimed synchrony otherwise?

Addressing the first issue, the proposed extended scheduler was realized to incorporate an empirically verified forward model that predicts an estimate of the gesture preparation time required by the robot prior to actual execution. For this, several approaches were implemented and evaluated as described in the following. Addressing the second issue, an on-line adjustment mechanism was integrated into the synchronization process for cross-modal adaptation within a chunk based on afferent sensory feedback. As before, scheduling, generation, and continuous synchronization of gesture and speech are flexibly conducted at run-time. **Figure 6.1** illustrates the 'planning-execution' procedure of the proposed improved scheduler; revised or extended features of the model are framed in red. The complete multimodal generation process within a multimodal chunk can be summarized as follows.

Phase 1: Planning

- **Speech preparation.** The planning phase begins with the phonological encoding of the designated speech output contained in the <specification> tag of a given MURML file. As described in Section 6.1.1, the MARY text-to-speech system establishes a complete list of phonemes and their respective durations during this process. Based on this list of phoneme durations, relevant speech timing information such as the onset time of the affiliate is determined. Speech output is subsequently generated in an audio file which is further processed in dependence on the position of the affiliate: if the affiliate is located at the beginning of the utterance, the audio file remains as it is; if the affiliate is located in the middle or at the end of the utterance, the sound file is split into two parts as follows. The first part contains the speech output to be uttered before the affiliate onset; the second part contains the speech affiliate and, if applicable, any subsequent remaining parts of speech.

- **Gesture preparation.** Based on the overt gesture form features specified in the <gesture> tag of the given MURML file, a suitable trajectory for the ACE body model is calculated, resulting in a movement plan. Meanwhile, the predictive forward model further described in Section 6.2 computes the

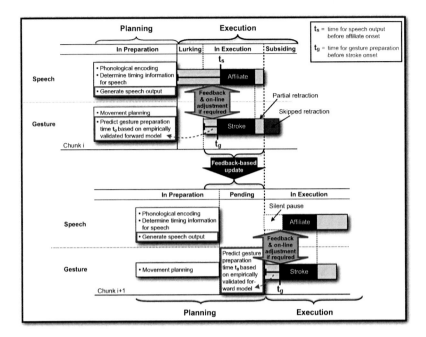

Figure 6.1: Model of the 'planning-execution' procedure of the extended scheduler; additions or changes in the model compared to the original scheduler are framed in red.

estimated execution time required by the robot for the gesture preparation phase before the stroke onset.

Based on a comparison of timing information for both modalities, start times for speech and gesture are determined so that temporal synchrony is achieved at the affiliate-stroke level. That is, if the duration of speech before the affiliate onset is longer than the estimated gesture preparation time before the stroke onset, then speech output starts first, otherwise gesture execution starts first.

Phase 2: Execution

- **Before the affiliate onset.** Speech-gesture production is initiated as scheduled in the final step of the planning phase. If there is speech output scheduled to precede the affiliate, the first sound file containing this part of speech is replayed.

While the robot performs the gesture preparation movement, variance between the target and its actual wrist position is constantly monitored utilizing afferent feedback from the WBM controller of the robot.

- **Ensuring synchrony between affiliate and stroke onset.** Once the robot's wrist has reached a position within a predefined range of the target location for the stroke onset, playback of the sound file containing the speech affiliate is triggered. If the predicted time was accurate, there should be no noticeable interruption in speech flow. If gesture preparation takes longer than scheduled, the feedback-based adaptation mechanism described in Section 6.3 becomes effective such that a pause is inserted before the speech affiliate.

In the following sections the implementation of the two major features extending the original ACE scheduler, namely the predictive forward model and the feedback-based adaptation mechanism, are described in more detail (see RQ5). Results obtained with these extensions are presented in each respective section.

6.2 Predictive Forward Model

Neurophysiological findings suggest that human motor control relies more on sensory predictions than on sensory feedback, as such feedback loops are considered too slow for efficient trajectory control given their inherent delay (Desmurget and Grafton, 2000; Wolpert and Flanagan, 2001). According to this view, the CNS performs predictions using an internal forward model that captures the causal relationship between motor commands and their sensory consequences (Kawato, 1999). Based on such model, the outcome of a rapid goal-directed movement is anticipated in real time using a so-called *efference copy* of the current motor commands (Von Holst and Mittelstaedt, 1950). For this, neural mechanisms of the CNS simulate the response of the motor system so that the outcome of a given motor command can be estimated prior to movement onset.

Drawing inspiration from this neurobiological perspective in light of the present research objective, the concept of internal forward models is of particular interest to the realization of the proposed scheduler for multimodal robot behaviors. Given the physical properties of the robot and its inability to move the arms with arbitrary speed, precise movement times of gesture execution are difficult to determine a priori. Importantly though, to successfully synchronize the robot's gesture with concurrent speech, the gesture preparation phase must be completed before the onset of the speech affiliate. Therefore, the execution time of this gesture phase in particular needs to be estimated as accurately as possible so that the behavior onset times of the two modalities can be adequately scheduled.

6.2.1 Overview of Possible Realization Approaches

As described in Section 5.3.2, the arm movement toward the stroke onset position in the preparation phase is generated as a linear trajectory in ACE. In the original framework, the according movement duration is estimated and modeled based on Fitts' Law (see Equation 5.1) which, however, was found to be inappropriate in the initial setup. To account for the modified motion generation constraints of the robotic platform and to achieve more accurate predictions of gesture execution time required for the preparation phase, the following three approaches were realized and evaluated:

1) **Fitting Fitts' Law.** In view of the original ACE implementation, the most straightforward approach aims at empirically determining a suitable value for the slope coefficient b of the Fitts' Law equation using a set of training data.

2) **Simple time estimation.** In an alternative but similarly straightforward approach a simple time estimation function representing the average velocity of the robot's arm is empirically approximated based on a set of training data.

3) **WBM-based trajectory simulation.** A more sophisticated approach comprises the internal simulation of the designated target trajectory for gesture preparation by using the robot-specific WBM controller.

Further potential approaches were identified, such as implementing a machine learning method (e.g., using neural networks) or integrating a look-up table containing previously collected timing information for a comprehensive set of trajectories. However, due to the limited time available for the present work, only the three approaches outlined above were realized, leaving these additional options for future investigation. Since the first two listed approaches are fairly straightforward and have already been illustrated in the previous chapter and in Figure 5.11, only the implementation of the third option is described in more detail in the next Section 6.2.2. Results assessing all three implemented approaches are subsequently reported and discussed in Section 6.2.3.

6.2.2 WBM-based Trajectory Simulation

The WBM software controlling the humanoid robot, as described in more detail in Section 4.2.2, provides the possibility for accelerated internal simulations of designated trajectories prior to the actual movement (Gienger et al., 2007). Although the identical WBM controller is used for both internal simulation and real robot control, the two processes are temporally decoupled. To simulate the robot's future state, the internal predictor iterates the robot model, however, ten times faster than real-time control. For this speed-up, longer sampling time intervals and less DOF than applied during real robot control are employed for the iterations (Sugiura et al., 2009). This way, the internal predictor model can simulate the desired trajectory much faster, requiring only negligible computation time. As a result, it offers a considerable option for the estimation of movement time required for the linear trajectory of the gesture preparation phase. The implementation of a forward model for ACE using the internal predictor model of the WBM controller is based on **Algorithm 1**.

Given the current position of the robot's arm ($x_{current}$) and the target position of the gesture stroke onset (x_{target}), the internal simulation iterates over the target trajectory $x(t)$ which is illustrated in **Figure 6.2a**. As long as the distance between current and target position is greater than a defined threshold value ε, in each iteration step 5 ms are added to the estimated trajectory time. Finally, the calculated time estimation value $t_{estimate}$ is returned to the ACE scheduler for multimodal utterance planning as depicted in Figure 6.1.

The outcome of the predicted trajectory time $t_{estimate}$ is influenced by a number of parameters for which appropriate values were to be determined as part of the present implementation and evaluation process. First, the threshold value ε providing the exit condition of the iteration loop needs to be specified so as to allow for an adequate number of iterations. Second, the speed of the simulated target

---

Algorithm 1 Trajectory time estimation

while $|x_{target} - x_{current}| > \varepsilon$ **do**
    iterate_trajectory();
    $t_{estimate}$ += 0.005;
end while
return $t_{estimate}$;

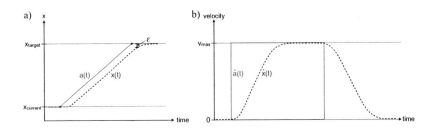

Figure 6.2: Step response of the WBM controller (adapted from Gienger et al., 2010): a) given two points $x_{current}$ and x_{target}, the attractor point for the controller is continuously shifted from the first to the latter as expressed by $a(t)$, resulting in the generation of the target trajectory $x(t)$; b) the velocity profile of $x(t)$ depends on the specified maximum velocity value v_{max} and on the relaxation time constant tmc which denotes the speed response of the controller and characterizes the slope of $\dot{x}(t)$.

trajectory depends on the preset maximum velocity value v_{max} (see **Figure 6.2b**): the greater the determined speed limit within the feasible range, the shorter the estimated execution time. Third, the relaxation time constant tmc affects the slope of the velocity profile of the trajectory so that a greater predefined speed response results in slower acceleration and thus a longer overall predicted trajectory time. Note that for real robot control, v_{max} and tmc are assigned the recommended values 0.8 m/sec and 0.2 sec respectively, resulting in smooth and non-overshooting trajectories; however, given the simplified iteration process of the internal simulation, these values may vary in the predictive model to allow for more accurate estimations.

Generally, due to the coarse nature of the simulation as well as the fact that the internal predictor model does not account for the robot's collision avoidance module, resulting predictions are potentially imprecise. Therefore, and in view of the three variable parameters listed above, the implemented model should also be fitted in a way that is similar to the other implemented options outlined in Section 6.2.1. The process that was undertaken to approximate the WBM-based forward model as well as the other two realized approaches is described in the following section together with a presentation and discussion of the results.

6.2.3   Technical Results and Discussion

To obtain an empirically validated forward model for the extended ACE scheduler, three approaches – WBM-based trajectory simulation, Fitts' Law, and simple time estimation – were implemented and fitted using a set of experimental training data. The data set comprised 60 trajectories of the right arm, each starting from the robot's default arm position and reaching to different targets within a selected range of typical gesture space. Each test trajectory was specified in a MURML utterance file in which the three positional values of the "HandLocation" slot (cf. Figure 4.5) were modified according to the combinations illustrated in **Figure 6.3**. The resulting set of targets specified a diverse range of both short-distance and long-distance trajectories; some of them (e.g., utterance no. 1) involved the potential to trigger the self-collision avoidance mechanism of the real-time robot controller, while others (e.g., utterance no. 60) were inherently self-collision-free.

Evaluation was based on comparisons with actual trajectory time values derived from the robot's performance of each test trajectory. Actual time values were measured using a threshold of 0.03 meters Euclidean distance between the robot's wrist and the target wrist position for the gesture stroke onset. That is, movement time for each trajectory was recorded until the robot's wrist entered the specified range around the target location. In order to assess the fitness and suitability of the three outlined forward models and to identify optimal values for the required parameters, each model was tested with a variety of parameter values as follows.

WBM-based Trajectory Simulation

- **Approximation threshold ε.** The following ten values were tested (in meters):
  $\varepsilon = \{0.005, 0.01, 0.015, 0.02, 0.025, 0.03, 0.035, 0.04, 0.045, 0.05\}$

- **Maximum velocity parameter v_{max}.** The following twelve values were tested (in meters per second; recommended: ≤ 0.8):
  $v_{max} = \{0.25, 0.3, 0.35, 0.4, 0.45, 0.5, 0.55, 0.6, 0.65, 0.7, 0.75, 0.8\}$

- **Relaxation time constant tmc.** The following eight values were tested (in seconds; recommended: ≥ 0.1):
  $tmc = \{0.15, 0.2, 0.25, 0.3, 0.35, 0.4, 0.45, 0.5\}$

The 960 resulting combinations were tested by comparing the predicted trajectory time value of each combination with the actual execution time required by the

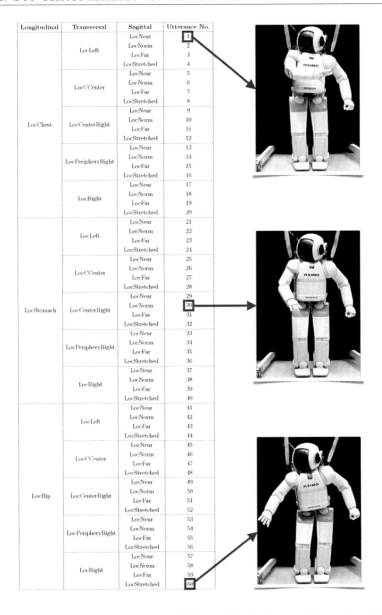

Longitudinal	Transversal	Sagittal	Utterance No.
LocChest	LocLeft	LocNear	1
		LocNorm	2
		LocFar	3
		LocStretched	4
	LocCCenter	LocNear	5
		LocNorm	6
		LocFar	7
		LocStretched	8
	LocCenterRight	LocNear	9
		LocNorm	10
		LocFar	11
		LocStretched	12
	LocPeripheryRight	LocNear	13
		LocNorm	14
		LocFar	15
		LocStretched	16
	LocRight	LocNear	17
		LocNorm	18
		LocFar	19
		LocStretched	20
LocStomach	LocLeft	LocNear	21
		LocNorm	22
		LocFar	23
		LocStretched	24
	LocCCenter	LocNear	25
		LocNorm	26
		LocFar	27
		LocStretched	28
	LocCenterRight	LocNear	29
		LocNorm	30
		LocFar	31
		LocStretched	32
	LocPeripheryRight	LocNear	33
		LocNorm	34
		LocFar	35
		LocStretched	36
	LocRight	LocNear	37
		LocNorm	38
		LocFar	39
		LocStretched	40
LocHip	LocLeft	LocNear	41
		LocNorm	42
		LocFar	43
		LocStretched	44
	LocCCenter	LocNear	45
		LocNorm	46
		LocFar	47
		LocStretched	48
	LocCenterRight	LocNear	49
		LocNorm	50
		LocFar	51
		LocStretched	52
	LocPeripheryRight	LocNear	53
		LocNorm	54
		LocFar	55
		LocStretched	56
	LocRight	LocNear	57
		LocNorm	58
		LocFar	59
		LocStretched	60

Figure 6.3: Test utterances used for the fitting of the implemented forward models.

real robot for each of the 60 test trajectories. Based on the smallest mean error across all 60 trajectories, the following values were identified as yielding the best combination: $\varepsilon = 0.015$, $v_{max} = 0.8$, $tmc = 0.2$. For the given training data, the mean prediction error was 0.1458 seconds with a minimum deviation of 0.0024 seconds for the best predicted trajectory time, and a maximum deviation of 0.6180 seconds for the poorest trajectory time prediction.

Fitts' Law

The forward model based on Fitt's Law was tested with the following 20 values for the slope coefficient $b = \{0.12, 0.14, 0.16, 0.18, 0.2, 0.22, 0.24, 0.26, 0.28, 0.3, 0.32, 0.34, 0.38, 0.4, 0.42, 0.44, 0.46, 0.48, 0.5\}$. Comparisons of mean error values for each of the 20 values across the 60 test trajectories promoted the coefficient value $b = 0.2$ as best choice. With a mean error of 0.1385 seconds, the Fitts' Law approach was found to perform slightly better than the WBM-based predictive model with the given values on the tested data set. The best trajectory time prediction yielded a minimum deviation of 0.0059 seconds, while the worst time prediction resulted in a maximum deviation of 0.5545 seconds.

Simple Time Estimation

The predictive model using a simple time estimation based on the assumed average velocity of the robot's arm movement was tested with the following 16 values (in meters per second): $avgVelocity = \{0.15, 0.16, 0.17, 0.18, 0.19, 0.20, 0.21, 0.22, 0.23, 0.24, 0.25, 0.26, 0.27, 0.28, 0.29, 0.30\}$. The value $avgVelocity = 0.28$ was identified as the best match for the given data set, resulting in a mean error of 0.2446 seconds which is substantially greater than in the other two approaches. With a minimum deviation of 0.0002 seconds more accurate prediction performance was found for the best case, while the maximum deviation value of 0.8942 seconds exceeded the worst case performances of the other two implemented approaches.

Discussion

The performances of the three realized approaches of predictive models using the best respective parameters as highlighted above are depicted in **Figures 6.4**, **6.5**, and **6.6**. The figures illustrate the gesture preparation times for each test utterance as predicted by the three forward models in comparison to the actual

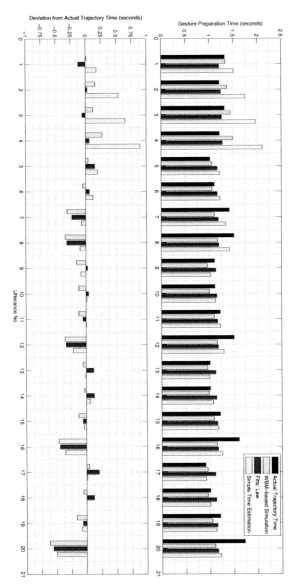

Figure 6.4: Comparison of performance of the three realized forward models for the test utterances no. 1–20.

141

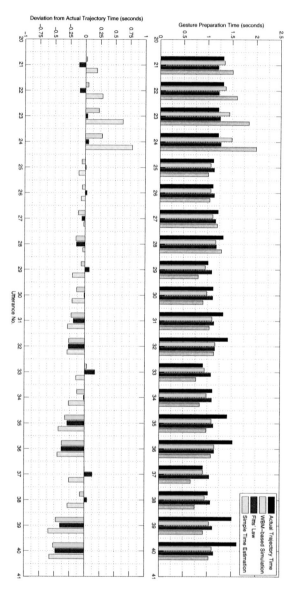

Figure 6.5: Comparison of performance of the three realized forward models for the test utterances no. 21–40.

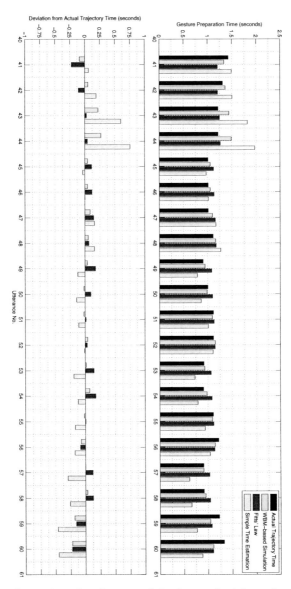

Figure 6.6: Comparison of performance of the three realized forward models for the test utterances no. 41–60.

trajectory time required by the real robot. In the left plot of each figure, the deviation between predicted and actual execution time of each test trajectory is visualized for the three strategies.

The deviation bars confirm what the calculated mean error values listed above suggest: the simple time estimation approach performs worst, especially when trajectory targets are located at a greater distance (e.g., utterances no. 3, 4, 23, 24). Moreover, with the selected average velocity parameter, the simple time estimation tends to overestimate the required gesture time more than the other two models (e.g., utterances no. 2, 3, 4, 23, 24, 43, 44).

Both the WBM-based simulation and the Fitts' Law approach show a general tendency to underestimate the actual trajectory time in most test cases. Furthermore, prediction errors and thereby deviations from actual trajectory time increase with greater distance to the target, regardless of the direction. For example, on average deviations are smaller for utterances no. 41 to 60 – all targeting the height of the hip which is located at a similar vertical level as the default position of the robot's arm – than for the other 40 test trajectories.

Generally, the prediction model based on the WBM controller and the forward model using Fitts' Law provide fairly similar performance qualities. Since other robotic platforms do not allow for the internal simulation of trajectories prior to execution, Fitts' Law therefore offers a viable alternative to the implemented WBM-based approach. This would facilitate a potential future transfer of the system onto other humanoid robots. With regard to the present work, however, the WBM-based forward model was utilized for the realization of the extended ACE scheduler. Since the simulation employs the same controller as used for subsequent generation, the WBM-based prediction model can better account for the actual path of the simulated trajectory with regard to the robot's physical properties, e.g., joint limits. In contrast, predictions using Fitts' Law are merely based on the calculated distance between start and target position, independent from their relative locations in the robot's gesture space.

As the trajectories used to identify suitable parameters for the tested models are not exhaustive, the WBM-based forward model may actually result in a smaller mean error value compared to the Fitts' Law approach if evaluated using a greater data set. Generally, given the number of adjustable parameters as well as their range of possible values, out of which only few were tested, there is ample room for improvement and further fine-tuning of the outlined models. Potentially, the incorporation of machine learning algorithms, e.g., by using neural networks as proposed in the list of possible approaches in Section 6.2.1, may lead to even more accurate forward models. These propositions, however, go beyond what could be realized in the time frame available for the present work and therefore remain points of future investigation.

Despite its limitations, the realized WBM-based forward model provides a more advanced – and with a mean error below 0.15 seconds generally acceptable – approach to gesture time planning compared to the original ACE scheduler. At the same time, the results suggest that a completely accurate prediction model is virtually impossible to achieve, thus emphasizing the lasting need for reactive feedback-based adaptation mechanisms during the gesture execution phase.

6.3 Feedback-Based Adaptation Mechanism

Despite the improved accuracy of timing estimation for the gesture preparation phase based on the implemented forward model, the actual timing of multimodal utterances might still deviate from the prediction during execution. Therefore, the second feature realized in the extended version of the ACE scheduler provides a feedback-based adaptation mechanism which allows for reactive cross-modal adjustment during the execution phase.

Generally, two types of real-time deviation from scheduled timing plans are possible: first, the forward model may *overestimate* the time required for gesture preparation, resulting in a premature gesture stroke onset; second, the predictive model may *underestimate* the time required for the gesture preparation phase, resulting in a delayed stroke onset. According to findings from human gesture research stating that the gesture stroke may precede but never follow the speech affiliate (e.g., McNeill, 1992), the first type of deviation appears less problematic than the second. Since the benchmark results of the WBM-based predictive model (visualized in Figures 6.4 to 6.6) document a maximal overestimation error of only 0.276 seconds for the tested data set, such cases may well be tolerated. In general, the implemented forward model was found to overestimate gesture preparation time by more than 0.1 seconds only in very rare instances (7 out of 60 cases), which further supports the decision to leave this type of prediction error unattended. In contrast, if the model exposed a pattern of frequent overestimation of gesture preparation time, the implementation of a pre-stroke hold phase would appear reasonable. However, given its inherent tendency to rather underestimate the time required by the robot, the present approach focuses on the more problematic type of prediction error with regard to gesture preparation time.

6.3.1 Implementation of Cross-Modal Adjustment

In his work, Kendon (2004) demonstrated that human speech may pause before the affiliate onset if the gesture preparation movement has not yet been completed, suggesting that the inserted pause ensures temporal synchrony of the gesture stroke with speech. Similarly, experiments conducted by Levelt et al. (1985) and de Ruiter (1998) showed that synchronization between the two modalities can indeed be achieved by the timing of speech adapting to the timing of gesture. As mentioned in Section 3.1.1, the original ACE scheduler does not account for such speech adaptation to gesture within a multimodal chunk, since the underlying processing model operates fully ballistically at intra-chunk level during the execution phase. When used on a robotic platform, this lacking adaptability is particularly problematic if gesture preparation time is significantly underestimated, in which case the delay of the gesture stroke onset may cause speech-gesture mistiming.

To account for the human ability to align speech to gesture also within a chunk, the ACE scheduler was modified to replace its open-loop execution mechanism with a more flexible, closed-loop approach. This is achieved by utilizing afferent feedback from the robot's WBM controller (cf. Figure 5.4) for reactive on-line adjustment of the multimodal synchronization process. Specifically, as outlined in the description of the remodeled execution phase in Section 6.1.3, cross-modal intra-chunk adaptation is achieved as follows. The sensory feedback received from the WBM controller provides information about the current wrist position of the robot's arm at each time step, which enables the scheduler to constantly check for variance between the target and actual position. This information is used as a 'triggering' mechanism, so that the speech affiliate is only uttered once the robot's hand has reached a position within the range of the stroke onset position.

If gesture preparation time is overestimated or correctly predicted, feedback about successful completion of the preparation phase is transmitted to the speech processing module before or at the very latest at the scheduled affiliate onset time. As a result, speech output is uttered according to the previously established generation plan without any audible disruption. If, however, gesture preparation time is underestimated, the described cross-modal processing dependency results in a deviation of the actual multimodal production process compared to the scheduled behavior plan. That is, if the robot's hand has not yet reached the stroke onset location by the predicted time, the reactive feedback mechanism intervenes in the speech production process: if the affiliate is located at the beginning of the utterance, it causes a delayed speech onset; if the affiliate is situated in the middle or at the end of speech, it results in an inserted speech pause right before the affiliate. To re-establish speech-gesture synchrony in the latter case, speech is paused until sensory feedback confirming the completion of the gesture preparation phase triggers its continuation. Such anticipation and reactive adaptation of the flow of speech to gesture timing complies with psycholinguistic models from the gesture literature promoting an interactive view of the relationship between speech and gesture (cf. Section 2.2.3).

Thus, the implemented feedback-based adaptation mechanism extends the ACE scheduler with a powerful means to achieve speech-gesture synchrony even on a more demanding robotic platform. This way, the extended scheduler allows for cross-modal adjustments not only between, but also within chunks.

6.3.2 Technical Results and Discussion

The following results were generated using the complete extended version of the ACE scheduler which incorporates both the empirically validated WBM-based predictive model and the reactive adaptation mechanism. In the ideal case, the movement time required for gesture preparation is estimated precisely enough during the planning phase to result in actual execution that is consistent with the scheduled multimodal plan. In such instance, the feedback-based adaptation mechanism would trigger the production of the speech affiliate just on time so that speech production is not notably interrupted. However, to demonstrate the cross-modal adjustment capability of the improved scheduler, examples were deliberately generated in which sensory feedback induced the insertion of pauses into the flow of speech. Such cases typically emerge when the gesture preparation time is considerably underestimated by the predictive model during planning.

As highlighted in Section 6.2.3, the forward model was found to produce the most prominent prediction errors of this type with increasing length of the anticipated gesture trajectory. Therefore, particularly distant locations were specified as targets for the gesture stroke onset in the following examples of multimodal robot behavior which were generated with the extended ACE scheduler. Two scenarios are demonstrated and discussed in the following: first, a multimodal utterance in which the speech affiliate is located at the beginning of the chunk, and second, one in which the affiliate is situated in the middle of the utterance.

Affiliate at Beginning of Chunk

Figure 6.7 depicts an example of a multimodal utterance generated with the extended ACE scheduler which illustrates the feedback-based adaptation mechanism effective at the beginning of the utterance. The upper plotted graph visualizes the velocity profile of the robot's right wrist during gesture execution; the lower graph plots the z-axis of the robot's wrist trajectory over time. Speech output is transcribed in temporal alignment to the generated gesture trajectory with words of the affiliate highlighted in red. The figure illustrates the underestimation of gesture preparation time by ∼0.4 seconds which results in a feedback-induced delay of the speech affiliate onset until the stroke onset position has been reached. In this way, despite the initially mistimed gesture preparation onset, temporal synchrony between the speech affiliate and the gesture stroke is ensured.

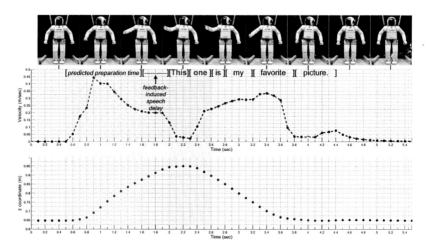

Figure 6.7: Multimodal utterance generated with the extended ACE scheduler demonstrating a feedback-induced speech delay prior to speech onset.

148

Affiliate in the Middle of Chunk

Figure 6.8 shows an example of a multimodal utterance generated with the extended ACE scheduler to illustrate the feedback-based adaptation mechanism effective in the middle of the chunk. As described in Section 6.1.3, the position of the affiliate yields the generation of two separate speech output files. Initially, speech and gesture commence according to their assigned onset times as scheduled in the planning phase. In this case, speech onset precedes the beginning of the gesture preparation phase, i.e., playback of the first audio file containing speech to be uttered before the affiliate onset is initiated first and is accompanied by gesture movements after ~0.72 seconds. As in the previous example, gesture preparation time is underestimated, resulting in a feedback-induced speech pause immediately preceding the affiliate onset. Once the preparation phase of the gesture has been completed, the feedback mechanism triggers the continuation of speech so that the affiliate and remaining speech output are replayed from the second audio file.

The two presented examples illustrate the functionality of the reactive cross-modal adaptation mechanism provided by the new ACE scheduler at the intra-chunk level. In this way, the extended scheduler not only provides cross-modal

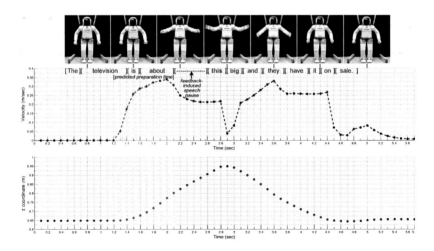

Figure 6.8: Multimodal utterance generated with the extended ACE scheduler demonstrating a feedback-induced speech pause within a chunk.

alignment between two successive chunks, but also within a single chunk. As a result, synchrony between robot gesture and speech can be ensured during the execution phase in spite of previous prediction errors made during the behavior planning and scheduling phase. This increased generation flexibility is a novelty in robotic platforms that aim at synchronized speech-gesture production and thus constitutes another major contribution of the present work.

6.4 Summary

This chapter focused on the synchronization of generated robot gestures with accompanying speech. In light of the special requirements and constraints imposed by the physical properties of the robot, the functionality of the original ACE scheduler was found to be insufficient for successful multimodal synchronization. Specifically, due to potentially mistimed behavior scheduling in the planning phase and the subsequent ballistic execution of single utterance chunks, the co-verbal gesture stroke onset was occasionally generated so that it followed the affiliate onset. This, however, conflicts with findings from human gesture research stating that the stroke generally precedes or, at the latest, begins at the onset of the affiliate in speech. Human gesturing behavior was further shown to allow for an adaptation of speech to gesture even within an intonation phrase if gesture preparation has not yet been completed prior to the affiliate onset (Kendon, 2004).

To optimize the synchronization of robot gesture and speech as envisaged by the present work, in **Section 6.1** the concept of an extended multimodal scheduler overcoming these conceptual shortcomings of the original ACE scheduler was presented. The proposed scheduler comprises two features that improve the synchronization process on the given robotic platform.

First, as described in **Section 6.2**, it was realized by incorporating an empirically verified forward model that predicts a more accurate estimate of the gesture preparation time required by the robot prior to actual execution. For this, three approaches – WBM-based trajectory simulation, based on Fitts' Law, and a simple time estimation – were implemented, fitted, and evaluated using a set of experimental test data. Eventually, the WBM-based approach was integrated into the final model, as it can better account for the robot's joint limits and the actual path of the simulated trajectory than the other two models.

The second feature of the extended scheduler was described in **Section 6.3**.

It incorporates an on-line adjustment mechanism into the synchronization process for cross-modal adaptation within a chunk based on sensory feedback from the robot. Specifically, information about its current wrist position is used to trigger the speech affiliate once the robot's hand has reached a position within the range of the stroke onset position, thereby ensuring temporal synchrony.

As a major contribution of the present work, the implementation of the proposed extended scheduler enables the humanoid robot to plan, generate, and continuously synchronize gesture and speech at run-time. With its empirically validated forward model and the possibility of cross-modal adaptation within a chunk, the extended scheduler represents the first closed-loop approach to speech-gesture generation for humanoid robots. Although there is certainly room for improvement with regard to both extended features, the presented scheduler already provides a more flexible and natural way to realize multimodal behavior for robots and other artificial communicators.

Part III: Empirical Evaluation

"The true delight is in the finding out
rather than in the knowing."
Isaac Asimov

Chapter 7

Empirical Evaluation - Study 1

Building on the technical work presented in the previous part of the thesis, this chapter introduces the empirical evaluation of speech and gesture generation with the humanoid robot. Note, as highlighted in Section 1.4 of the introduction, the primary purpose of the studies presented in the following was not a technical evaluation in the sense of testing or benchmarking different implementations of the framework. Rather, the aim was to utilize the realized system as a tool to investigate more general research questions regarding the acceptance and evaluation of robot gestures in human-robot interaction (HRI).

As outlined in the review of related gesture-based HRI studies in Section 3.2.2, the evaluation of co-verbal robot gesture, especially with regard to effects, perception, and acceptance thereof, is still in the early stages. With only few empirical studies and results reported in the relevant literature so far, there is still ample scope for further research. The second major objective of the present work is thus to contribute a set of new empirical findings about the effects that communicative robot gesture may have on human interaction partners. For this purpose, two between-subjects experimental studies were conducted using the humanoid robot and the realized multimodal action generation framework.

The first study is presented in this chapter, which is organized as follows. In Section 7.1 the methodology of the conducted study is described, including the experimental design and procedure, the proposed hypotheses, the dependent measures used, and information on the participants of the study. Section 7.2 then provides exploratory results based on the participants' evaluation of the behavior displayed by the robot as well as based on their own performance during and after the interaction.

7.1 Method

To gain a deeper understanding of how communicative robot gesture may impact and shape human experience in HRI, a suitable interaction scenario was designed and benchmarks for the evaluation were identified. The study scenario comprised a joint task that was to be performed by a human participant in collaboration with the humanoid robot. The main motivation for choosing a task-based interaction was to realize a largely controllable yet meaningful interaction which would allow for a measurable comparison of participants' reported experiences. In the given task, the robot referred to various objects by utilizing either unimodal (speech only) or multimodal (speech and gesture) utterances, based on which the participant was expected to perceive, interpret, and perform an according action.

7.1.1 Experimental Design

The experiment was set in a simulated kitchen environment within the robot lab as depicted in **Figure 7.1a**. The humanoid robot served as a household assistant. Participants were told that they were helping a friend who was moving house. They were tasked with emptying a cardboard box with kitchen items, each of which had to be placed in its designated location. The box contained nine kitchen items whose storage placement is not typically known a priori (unlike plates, for example, which are usually piled on top of each other).[1] Specifically, they comprised a thermos flask, a sieve, a ladle, a vase, an eggcup, two differently shaped chopping boards and two differently sized bowls. The cardboard box containing the kitchen items used in the experiment is displayed in **Figure 7.2**.

The objects were to be removed from the box and arranged in a pair of kitchen cupboards (upper and lower cupboard with two drawers, see Figure 7.1a). For this, the participant was allowed to move freely in the area in front of the robot, typically walking between the cardboard box with items and the kitchen cupboards. Given the participant's non-familiarity with the friend's kitchen environment, the robot was made to assist with the task by providing information on where to put the respective kitchenware. For the case that the participant did not understand

---

[1]This design choice was made to prevent participants from deducing the location of the item from common conventions rather than by strictly following the robot's instructions.

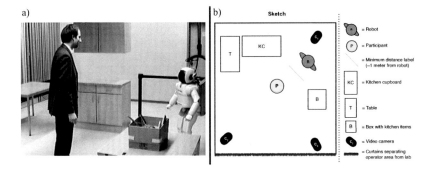

Figure 7.1: The experimental setting in the lab (adapted from Salem et al., 2011a): a) participant of Study 1 interacting with the robot in a simulated kitchen environment; b) sketch of the experimental area in the lab.

Figure 7.2: Cardboard box containing kitchen items used in the experimental study (reprinted from Salem et al., 2012).

where the item had to be stored, a table situated beside the kitchen cupboard was provided for alternative placement.[2] The entire interaction was filmed by three video cameras from different angles of the experimental area. A sketch of the experimental setting is illustrated in **Figure 7.1b**.

Conditions

The non-verbal behaviors that were displayed by the humanoid robot were manipulated in three experimental conditions:

1) In the **unimodal** (*speech-only*) condition, the robot presented the participant with a set of nine verbal instructions to explain where each object should be placed. The robot did not move its body during the whole interaction; no gesture or gaze behaviors were performed.

2) In the **congruent multimodal** (*speech-gesture*) condition, the robot presented the participant with the identical set of nine verbal instructions used in the unimodal condition. In addition, they were accompanied by a total of 21 corresponding gestures explaining where each object should be placed. Speech and gesture were semantically matching (e.g., the robot said "upper cupboard" and pointed up). Simple gaze behavior supporting the hand and arm gestures (e.g., looking right when pointing right) was displayed during the interaction.

3) In the **incongruent multimodal** (*speech-gesture*) condition, the robot presented the participant with the identical set of nine verbal instructions used in the unimodal condition. Again, in addition, they were accompanied by a total of 21 gestures. However, out of these only ten gestures (47.6 %) semantically matched the respective verbal instruction, while the remaining eleven gestures (52.4 %) were semantically non-matching (e.g., the robot occasionally said "upper cupboard" but pointed downwards). Simple gaze behavior supporting the hand and arm gestures (e.g., looking right when pointing right) was displayed during the interaction.

The incongruent multimodal condition was designed to decrease the reliability and task-related usefulness of the robot's gestures. In other words, participants in

---

[2]Participants were explicitly advised not to guess the location of the item in such case so that their performance could be correctly evaluated afterwards.

this group were unlikely to evaluate the use of the additional gesture modality solely based on its helpfulness in solving the given task. In this way, the third condition provided a means to investigate whether the mere existence of non-verbal behavior, even if not particularly appropriate, would still positively affect the evaluation of the robot compared to the speech-only condition. The choice to combine semantically non-matching gestures with matching ones in this condition was made to avoid a complete loss of the robot's credibility after a few utterances.

Verbal Utterances

In order to keep the task solvable under all three conditions, the spoken utterances delivered by the robot were designed in a self-sufficient way. That is, the gestures used in the multimodal conditions contained additional illustrative information which was not indispensable to solving the task. Each instruction presented by the robot typically consisted of two or three continuously connected unimodal or multimodal chunks (see Definition 2 in Section 2.2.1). The verbal utterance chunks were based on the following syntax:

- **Two-chunk utterance:**
  ```
  <Please take the [object]>
  <and place it [position + location].>
  ```
  Example: *Please take the thermos flask and place it on the right side of the upper cupboard.*

- **Three-chunk utterance:**
  ```
  <Please take the [object],>
  <then open the [location]>
  <and place it [position].>
  ```
  Example: *Please take the eggcup, then open the right drawer and place it inside.*

Gestures

In the multimodal conditions, the robot used three different types of gesture along with speech to indicate the designated placement of each item:

- **Iconic gestures**, e.g., to illustrate the shape or size of objects

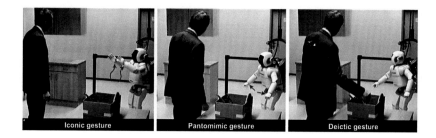

| Iconic gesture | Pantomimic gesture | Deictic gesture |

Figure 7.3: Examples of the three types of gesture performed by the robot during interaction in the multimodal conditions (left to right; blue arrows indicating the movement trajectory and direction of dynamic gestures): *iconic gesture* illustrating the shape of the vase; *pantomimic gesture* conveying the act of opening the lower cupboard; *deictic gesture* pointing at the designated location (adapted from Salem et al., 2011c).

- **Pantomimic gestures**, e.g., hand movement performed when opening cupboard doors or using a ladle

- **Deictic gestures**, e.g., to indicate positions and locations

Examples of the three gesture types are illustrated in **Figure 7.3**. A complete list of instructions presented by the robot in the different experimental conditions during interaction is provided by **Table 7.1**.

Robot Control and Behavior

For the control of the robot during the experimental study the realized speech-gesture generation framework described in Part II of this thesis was used. Since the implementation of the extended multimodal scheduler had not been completed by the time this study was conducted, the original ACE scheduler was employed for the generation of multimodal behavior. Besides the preliminary adjustment of the function estimating gesture execution time as described in Section 6.1.2, the utterances used for the experiment were specifically designed to provide reasonable synchrony between the two modalities. Speech output produced by the robot was identical across conditions and was generated using the text-to-speech synthesis system MARY (see Section 6.1.1) set to a neutral German voice.

To ensure minimal variability in the experimental procedure, the robot was partly controlled using a Wizard-of-Oz technique during the study. The experiment

Table 7.1: Instructions presented by the robot during interaction: the respective item being referred to in the verbal utterance (translated into English) is printed in bold letters; gestures performed by the robot are depicted for each condition, with blue arrows indicating the movement trajectory and direction of dynamic gestures; non-matching gestures of the incongruent multimodal condition are framed in red.

Speech	Gesture		
	Unimodal	Congruent Multimodal	Incongruent Multimodal
Please take the **thermos flask** and place it on the right side of the upper cupboard.			
Please take the **rectangular chopping board**, then open the lower cupboard and place it in the middle.			
Please take the **large bowl** and place it on the left side of the upper cupboard.			
Please take the **egg cup**, then open the right drawer and place it inside.			
Please take the **vase** and place it on the left side of the lower cupboard.	(no gesture)		
Please take the **sieve**, then open the upper cupboard and place it in the middle.			
Please take the **small bowl** and place it inside the left drawer.			
Please take the **soup ladle** and place it on the right side of the lower cupboard.			
Please take the **round chopping board** and place it in the middle onto the lower cupboard.			

161

room was partitioned with a curtain such that the robot and kitchen environment were located at one end and the wizard operating the control computer was located at the other end, outside the participant's field of view. The experimenter initiated the robot's interaction behavior from a fixed sequence of predetermined utterances. Once triggered, a given utterance was generated autonomously at run-time using the implemented ACE-based action generation framework. The ordering of the utterance sequence remained identical across conditions and experimental runs.

The robot delivered each two-chunk or three-chunk instructional utterance as a singular one-shot expression without any significant breaks in the delivery process. Successive chunks indicating object, position and location were delivered contiguously in the manner of natural speech. Moreover, in the two co-verbal gesture conditions, spoken utterances were accompanied by respective gestures. Participants were instructed to indicate when they had finished placing an item and were ready for the subsequent instruction by saying "next".

7.1.2 Experimental Procedure

Participants were tested individually. First, they received a description of the scenario as well as experimental instructions in written form to read outside the experiment room. They were then brought into the robot lab where the experimenter verbally reiterated the task instructions to ensure the participants' familiarity. Subsequently, participants were given the opportunity to ask any clarifying questions before the experimenter left the participant to begin the interaction with the robot.

At the beginning of the experiment, the robot greeted the participant and introduced the task before commencing with the actual instruction part. The robot then presented the participant with individual utterances as described in the experimental design, each of which was triggered by the experimenter observing and controlling the interaction from the adjacent room. The participant then followed the uttered instruction and, ideally, placed each item into its correct location. As explained in the briefing prior to the experimental task, participants were requested to place the object on a table adjacent to the kitchen cupboard if unsure about the designated location of the item, rather than trying to guess its final position. At the end of the interaction, the robot thanked the participant for helping and bid them farewell. Participants interacted with the robot for approximately five minutes. In the unimodal (speech-only) condition, all utterances including the greeting and farewell were presented verbally; in the multimodal (speech-gesture) conditions, all utterances including the greeting and farewell were accompanied by co-verbal gestures.

After interacting with the robot, participants were led out of the lab where they were asked to complete a post-experiment questionnaire to evaluate the robot and the interaction experience. Upon completion of the questionnaire, participants were carefully debriefed about the purpose of the experiment and received a chocolate bar as a thank-you before being dismissed.

7.1.3 Hypotheses

Based on findings from gesture research in human-human as well as human-agent interaction, the following hypotheses for gesture-based human-robot interaction were developed:

H1: Participants who receive multimodal instructions from the robot (i.e., using speech and either congruent or incongruent gesture) will evaluate the robot more positively than those who receive unimodal information from the robot (i.e., using only speech).

H2: Participants who are presented with congruent multimodal instructions by the robot will perform better at the task than those who are presented with unimodal or incongruent multimodal information by the robot.

H3: Participants who receive multimodal instructions from the robot will take up more information about the objects handled during the interaction than those who are presented with unimodal information only.

7.1.4 Dependent Measures

Participants were asked to report their interaction experience with the robot and rate its perceived behavior based on several items in the post-experimental questionnaire. In addition, collected video data was analyzed to evaluate participants' performance during the task. Data analysis focused on four main aspects.

- *Quality of presentation* was measured using six questionnaire items based on five-point Likert scales as listed in **Table 7.2**.

- *Perception of the robot* was assessed using seven questionnaire items based on five-point Likert scales as presented in **Table 7.3**

- *Task-related performance of participants* was derived from the error rate, i.e., the number of objects that were not correctly placed by each participant during the experimental task, and using a questionnaire item based on a five-point Likert scale as shown in **Table 7.4**.

- *Information uptake* was derived from participants' recall rate of the first and last object which they had moved during the experimental task and was evaluated using the two open-ended questions from the questionnaire listed in **Table 7.5**.

Table 7.2: Dependent measures, respective questionnaire items and scales used to evaluate the *quality of presentation*:

Measure:	Questionnaire Item:	Scale:
Gesture Quantity	"The amount of gestures performed by the robot were..."	1 = too few, 5 = too many
Gesture Speed	"The execution of gestures was..."	1 = too slow, 5 = too fast
Gesture Fluidity	"The execution of hand and arm movements was fluid."	1 = strongly disagree, 5 = strongly agree
Speech-Gesture Content	"The robot's speech and gesture were semantically matching."	1 = strongly disagree, 5 = strongly agree
Speech-Gesture Timing	"The robot's speech and gesture were well synchronized."	1 = strongly disagree, 5 = strongly agree
Naturalness	"The combined use of speech and gesture appeared..."	1 = artificial, 5 = natural

Table 7.3: Dependent measures, respective questionnaire items and scales used to evaluate the *perception of the robot*:

Measure:	Questionnaire Item:	Scale:
sympathetic		
lively		
active		
engaged	"Please assess to which extent	1 = strongly disagree,
friendly	the following characteristics	5 = strongly agree
communicative	apply to the robot: [*measure*]"	
fun-loving		

Table 7.4: Dependent measure, respective questionnaire item and scale used to evaluate the *task-related performance of participants*:

Measure:	Questionnaire Item:	Scale:
Competence Self-Rating	"How competent were you in solving the task?"	1 = not at all, 5 = very

Table 7.5: Dependent measures and corresponding questionnaire item used to evaluate the *information uptake*:

Measure:	Questionnaire Item:
First Object Recall	"Which was the first item you had to put away?"
Last Object Recall	"Which was the last item you had to put away?"

7.1.5 Participation

A total of 60 people (30 female, 30 male) participated in the experiment, ranging in age from 20 to 62 years ($M = 31.12$, $SD = 10.21$). All participants were native German speakers who were recruited at Bielefeld University and had never before participated in a study involving robots. Based on five-point Likert scale ratings (1 = very little, 5 = very much), participants were identified as having negligible experience with robots ($M = 1.22$, $SD = 0.45$), while their computer and technology know-how was moderate ($M = 3.72$, $SD = 0.90$). Participants were randomly assigned to one of the three experimental conditions (i.e., between-subjects design with 20 participants per condition), while maintaining gender- and age-balanced distributions.

7.2 Results

7.2.1 Quality of Presentation

The perceived quality of the instructions presented by the robot was investigated with regard to gesture and its combined use with speech. Mean values and standard deviations are summarized in **Table 7.6**. Note that for the unimodal condition only gesture quantity was measured, as participants in this experimental group were not presented with any non-verbal behavior by the robot and thus could not rate the quality of the robot's gestures.

With regard to *gesture quantity*, the overall mean value for the two gesture conditions was $M = 2.90$ ($SD = 0.59$). This means, on average participants were quite satisfied with the gesture rate. For the unimodal condition, participants rated gesture quantity as rather low ($M = 1.90$, $SD = 0.99$), which can be attributed to the lack of non-verbal behavior displayed by the robot. Analysis of variance (ANOVA) comparing the three experimental groups with regard to ratings of *gesture quantity* showed a significant effect of condition ($F(2,47) = 8.93$, $p = .001$). Pairwise comparisons with Tukey's post-hoc test confirmed significantly higher ratings in both the congruent and incongruent multimodal groups when compared to the unimodal group with $p = .004$ and $p < .001$ respectively. No significant differences were found between the two gesture conditions. When combining the results of the two multimodal conditions, the remaining five attributes measuring presentation quality were found to have the following overall mean values (standard

Table 7.6: Mean values of dependent measures rating the *quality of presentation* in the three conditions (standard deviations in parentheses).

	Condition		
Measure	Unimodal	Congr. Multimodal	Incongr. Multimodal
Gesture Quantity	1.90 (0.99)	2.80 (0.62)	3.00 (0.56)
Gesture Speed		2.85 (0.37)	2.95 (0.22)
Gesture Fluidity		3.25 (0.97)	3.95 (1.05)
Speech-Gesture Content		3.65 (1.04)	3.30 (1.26)
Speech-Gesture Timing		3.90 (0.79)	4.05 (1.10)
Naturalness		3.20 (1.06)	3.30 (1.13)

deviation in parentheses). Ratings of *gesture speed* yielded $M = 2.90$ ($SD = 0.30$), indicating suitable speed which was neither considered too slow nor too fast. *Gesture fluidity* yielded $M = 3.60$ ($SD = 1.06$) across the two multimodal conditions, that is, smoothness of the robot's gestures was perceived as acceptable but not completely fluid. Semantic matching of speech and gesture (*speech-gesture content*) yielded $M = 3.48$ ($SD = 1.14$), with higher ratings in the congruent than in the incongruent multimodal condition. Temporal matching of speech and gesture (*speech-gesture timing*) was perceived as fairly appropriate with $M = 3.97$ ($SD = 0.95$). Finally, perceived *naturalness* of the robot's combined use of speech and gesture was predominantly rated with values in the middle range of the scale, yielding $M = 3.25$ ($SD = 1.08$).

Additional comparisons of the two multimodal groups using independent-samples t-tests with 95 % confidence intervals showed a significant effect only for the *gesture fluidity* measure ($t(38) = -2.19$, $p = .034$). That is, gestures were perceived as significantly more fluid in the incongruent multimodal condition than in the congruent multimodal condition.

To summarize, the results indicate that participants were generally satisfied with the quality of co-verbal gestures performed by the robot with mean values above average on the 'strongly disagree – strongly agree' scales.

7.2.2 Perception of the Robot

Participants' perception of the humanoid robot was measured using the seven characteristics listed in Table 7.3. To test the first hypothesis (H1), independent-samples t-tests with 95 % confidence intervals were conducted as follows. First, questionnaire data from the unimodal condition were compared with data from the congruent multimodal condition. Second, data from the unimodal condition were compared with data from the incongruent multimodal condition. Mean values of the dependent measures reflecting participants' *perception of the robot* in the three different conditions are listed together with their standard deviation values in **Table 7.7** and are visualized in **Figure 7.4**. Items resulting in statistically significant effects when comparing the respective multimodal gesture condition with the unimodal speech-only condition are marked by asterisks (*).

Comparing the unimodal condition with the congruent multimodal condition, the four characteristics *sympathetic* ($t(38)$ = -1.90, p = .033), *lively* ($t(38)$ = -2.09, p = .022), *active* ($t(38)$ = -2.70, p = .005), and *fun-loving* ($t(38)$ = -2.12, p = .021) were found to be rated significantly higher in the congruent gesture condition than in the unimodal condition using speech only.

Table 7.7: Mean values of the dependent measures reflecting participants' *perception of the robot* (standard deviations in parentheses); $^{+}$ = p < .10, * = p < .05, ** = p < .01, *** = p < .001.

| | Condition | | |
Measure	Unimodal	Congr. Multimodal	Incongr. Multimodal
sympathetic	3.60 (1.05)	4.20 (0.95)*	4.15 (1.09)$^{+}$
lively	2.52 (0.84)	3.12 (0.97)*	3.32 (0.76)**
active	2.35 (0.88)	3.20 (1.11)**	3.45 (0.76)***
engaged	3.25 (1.29)	3.60 (1.35)	4.15 (0.88)**
friendly	4.15 (1.04)	4.35 (1.31)	4.60 (0.68)$^{+}$
communicative	3.00 (1.08)	3.15 (1.31)	3.60 (1.05)*
fun-loving	1.95 (0.83)	2.65 (1.23)*	2.70 (1.30)*

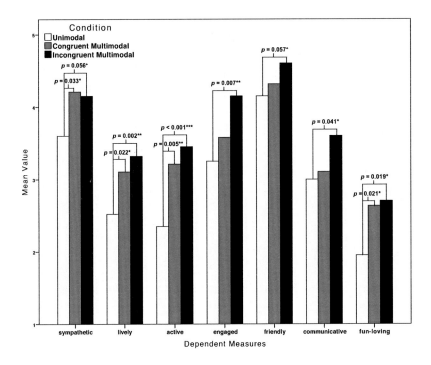

Figure 7.4: Bar chart visualizing the mean ratings and significant effects for the dependent variables measuring participants' *perception of the robot*; $^{+}$ = p < .10, * = p < .05, ** = p < .01, *** = p < .001.

When comparing the unimodal condition with the incongruent multimodal condition, the five characteristics *lively* (t(38) = -3.17, p = .002), *active* (t(38) = -4.25, p < .001), *engaged* (t(38) = -2.58, p = .007), *communicative* (t(38) = -1.79, p = .041), and *fun-loving* (t(32.16) = -2.18, p = .019) were found to be rated significantly higher in the multimodal condition. In addition, comparing the ratings for the characteristics *sympathetic* (t(38) = -1.63, p = .056) and *friendly* (t(38) = -1.62, p = .057) yielded a marginally significant effect, with higher mean values in the incongruent multimodal condition.

On average, in comparison to the the unimodal condition, all qualities were rated higher, i.e., more positively, in the two multimodal gesture conditions. These results support hypothesis H1 and suggest that the inclusion of gestural behavior casts the robot in a more positive light than in the speech-only condition.

An additional comparison of the two multimodal conditions, i.e., congruent vs. incongruent multimodal, showed no significant effect of experimental condition. However, with the exception of the measure *sympathetic*, analyses indicated a general trend towards higher mean values in the incongruent multimodal condition.

7.2.3 Task-Related Performance of Participants

Participants' task-related performance was measured in two ways: first, *subjective assessment* was measured using a questionnaire item asking participants to assess their own performance (see Table 7.4); second, *objective assessment* was derived from the task-related error rate, i.e., the number of objects that were not correctly placed during the experimental HRI task. This objective measure was also used to test the second hypothesis (H2).

Results of *subjective assessment* ratings are presented in **Table 7.8**. Generally, participants rated their own competence as rather high, with mean values between 4.55 and 4.60 in all three groups. No effect of experimental condition was found.

Results of *objective assessment* ratings are listed in **Table 7.9** and are illustrated in **Figure 7.5**. With an average error rate of 5.56 % across all nine kitchen objects, participants in the unimodal group made more task-related errors than participants in the two multimodal groups. The average error rate was further found to be slightly higher in the incongruent multimodal condition (1.11 %) than in the congruent multimodal condition (0.56 %). Although not yielding a significant effect, this overall trend is in line with hypothesis H2.

Finally, Spearman's correlation analysis showed a significant negative correlation between *objective* and *subjective* assessment measures ($r = -.31$, $p = .017$).

Table 7.8: Mean values of the measure indicating participants' *subjective assessment* (standard deviations in parentheses).

Measure	Unimodal	Condition Congr. Multimodal	Incongr. Multimodal
Competence Self-Rating	4.55 (0.69)	4.60 (0.60)	4.60 (0.50)

Table 7.9: Average error rates per object and group used to measure participants' *objective assessment*; asterisks (*) indicate items which were described in utterances with non-matching gestures in the incongruent multimodal condition.

	Condition		
Object	Unimodal	Congr. Multimodal	Incongr. Multimodal
Thermos Flask	5.0 %	0 %	0 %
Rect. Chopping Board*	15.0 %	0 %	0 %
Large Bowl*	15.0 %	0 %	0 %
Egg Cup*	0 %	0 %	5.0 %
Vase	0 %	0 %	0 %
Sieve*	10.0 %	0 %	0 %
Small Bowl*	0 %	0 %	0 %
Soup Ladle*	0 %	0 %	0 %
Round Chopping Board	5.0 %	5.0 %	5.0 %
Average Error per Group	5.56 %	0.56 %	1.11 %

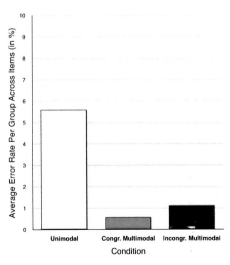

Figure 7.5: Bar chart visualizing participants' *objective assessment* based on the average error rate per group across all nine objects handled in the experimental task.

171

That is, the more errors participants made in the experimental task, the lower they actually rated their own competence afterwards.

7.2.4 Information Uptake

To test the third hypothesis (H3), participants' *information uptake* was measured based on two open-ended questions from the post-experimental questionnaire (see Table 7.5). Specifically, participants were asked to recall the first and last object which they had to move during the experimental task. As the respective questions were located at the end of the questionnaire which, for consistency, had to be filled out in the given order, participants could typically only answer them after more than ten minutes had elapsed since actually completing the experimental task. Note that in the incongruent multimodal condition, both the first and last object were accompanied by matching gestures, i.e., instructions delivered by the robot with respect to these two items were identical in both multimodal groups (see Table 7.1).

Average recall rates for each experimental group regarding the first and last object are presented in **Table 7.10** and are visualized in **Figure 7.6**. As predicted, recall rates were higher in the two multimodal conditions compared to the unimodal condition. To evaluate the observed differences, a Kruskal-Wallis test was conducted. Despite the clear trend, no statistically significant differences between the experimental groups were found when comparing the recall rates for either the first object or the second object alone. However, when comparing the percentage of participants in each group who were able to recall both, a significant effect of condition was found, $\chi^2(2) = 7.01$, $p = .030$. To follow up on this finding, pairwise comparisons of the three groups were conducted using Mann-Whitney tests with a Bonferroni correction resulting in a cut-off significance level of $p = .0167$ (i.e., .05 / 3). The post-hoc test showed a significant difference between the unimodal group and the congruent multimodal group ($U = 120.00$, $p = .006$, $r = -.40$). That is, participants who received instructions with congruent co-verbal gestures from the robot were significantly better at recalling both the first and last object handled during the HRI task. These results as well as the overall observed trend support hypothesis H3.

Table 7.10: Participants' *information uptake* per group based on the average recall rate regarding the first and last object moved during the experimental task.

	Condition		
Measure	Unimodal	Congr. Multimodal	Incongr. Multimodal
First Object Recall	50 %	75 %	70 %
Last Object Recall	65 %	85 %	85 %
First & Last Object Recall	35 %	75 %	64 %

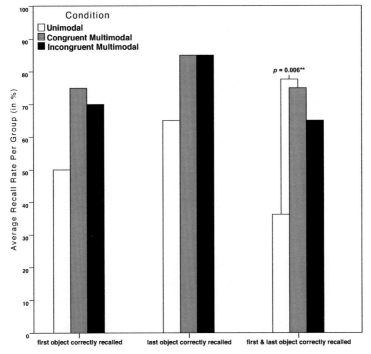

Figure 7.6: Bar chart visualizing participants' *information uptake* per group based on the average recall rate regarding the first and last object moved during the experiment.

7.3 Summary and Discussion

In this chapter, the first of two experimental studies investigating the effect of robot gesture was presented. To assess how representational hand gestures performed by the humanoid robot may impact human interaction partners, a task-related HRI scenario was designed based on the methodology described in **Section 7.1**. Participants were allocated the task to empty a cardboard box with kitchenware, while the robot assisted them by providing information about the storage location of each item. The non-verbal behavior of the robot was manipulated in three experimental conditions: (1) *unimodal* (only speech, no gesture), (2) *congruent multimodal* (speech and matching gesture), and (3) *incongruent multimodal* (speech and partly non-matching gesture). Hypotheses proposed in Section 7.1.3 predicted a positive effect of robot gesture on dependent variables measuring participants' *perception of the robot* and *information uptake*. In addition, a positive effect of congruent gesture was hypothesized with regard to participants' *task-related performance*.

Empirical results were reported in **Section 7.2**. Main findings supporting the hypotheses are interpreted and discussed in more detail in the following.

Perception of the Robot

In a first exploratory investigation of participants' perception and evaluation of the robot with a dependence on its non-verbal behavior, seven questionnaire items were analyzed to measure the attribution of various characteristics to the robot. The results support the first hypothesis (H1) which predicted a more positive assessment of the robot in the two multimodal conditions compared to the unimodal condition.

The significantly higher ratings of the characteristics *lively* and *active* in the two multimodal conditions can possibly be attributed to the robot's gestural movements, since the robot appeared comparatively stiff in the speech-only condition. The ratings of the characteristics *sympathetic, engaged, communicative,* and *fun-loving* further suggest that human-like non-verbal behaviors including gesture actually trigger a more positive response within the human participant. This was also found to be true for hand and arm gestures that did not semantically match the information content conveyed via speech. These findings imply that a humanoid robot that generates gestures – even if in part they are semantically

174

'incorrect' – is still more favorable than one that performs no gestures at all. Interestingly, on average the robot was evaluated as more *lively, active, engaged, friendly, communicative,* and *fun-loving* in the incongruent multimodal condition than in the congruent condition. This suggests that a robot's non-verbal communicative behavior can potentially trigger a more positive response within the human participant when it is not 'perfect'. Since this finding is rather surprising and exceeds the original hypothesis, implications from this study were further elucidated in the subsequent study to test whether this observation was replicable.

Overall, the results demonstrate that co-verbal gestures performed by a humanoid robot lead to an enhanced HRI experience, i.e., the robot is generally rated more positively when it displays non-verbal behaviors (see RQ6). These findings support the present approach of endowing social robots with communicative gestural behavior.

Task-Related Performance of Participants

Analysis of objective data measuring participants' performance in the experimental task, i.e., the number of objects that were not correctly placed, revealed a trend in favor of the second hypothesis (H2), but no significant effect was found.

On average, participants in the unimodal group made more mistakes during the task than participants in either of the multimodal groups. Surprisingly though, despite the partly contradictory information conveyed in the two modalities, participants in the incongruent multimodal group made a lot less – and generally only very few (1.11 %) – errors than participants in the speech-only condition. This observation gives reason to believe that participants in this group paid less attention to the robot's gestural behavior, perhaps even failing to notice any incongruity. In fact, this theory may explain the relatively high mean value of the *speech-gesture content* measure (see Table 7.6) which reflects the extent to which the participants perceived the robot's speech and gesture to be semantically matching. In terms of this measure, no significant difference was found between the participant ratings of the two multimodal groups. In addition, the above assumption may further account for why the characteristics measuring participants' perception of the robot were rated higher in the incongruent condition than in the congruent condition.

Generally, participants in all three experimental groups were often observed to focus more on solving the given task than on the robot giving the instructions. That is, once the robot had named the next object to be put away, many participants immediately walked towards the box of items in an attempt to retrieve it. During this period, the robot would still be delivering the following utterance chunk(s) stating the designated location of the object. This part of the utterance, however, comprised the majority of non-matching information in the incongruent condition.

The behavior displayed by the participants of this study is actually in line with findings from human gesture research showing that addressees rarely gaze directly at the speaker's gesture; instead, they typically spend as much as 90 to 95 % of the total viewing time fixating on the speaker's face (Gullberg and Holmqvist, 1999). Generally, participants' attentional behavior during human-robot interaction in this study can be viewed as a positive finding, as it indicates that they interacted in a fairly natural way. However, it is believed that it may have lessened their ability to consciously assess the robot's behavior. This issue was addressed in the second study which is described in Chapter 8.

Information Uptake

The third hypothesis (H3) tested in this study predicted higher recall rates from participants in the multimodal conditions regarding the first and last object handled during the experimental task. Inspiration for this hypothesis was drawn from studies in Cognitive Psychology showing that participants who are presented with a list of items generally tend to recall the first and last items better than those that are in the middle of the list. This is referred to as the *serial position effect* (Murdock, 1962); the tendency to recall earlier items has been termed the *primacy effect*, the tendency to recall later items is called the *recency effect*.

H3 was further based on the assumption that the recall of the two requested objects would be facilitated if the verbal stimulus produced by the robot was complemented by gesture-based imagery.[3] Such imagery was provided by the robot in the form of a pantomimic gesture illustrating the use of the first item (i.e., the thermos flask) and an iconic gesture depicting the shape of the last item (i.e., the round chopping board).

Results from the study were in line with this hypothesis, showing that participants in both multimodal conditions were better at recalling the first or last object

---

[3]Shams and Seitz (2008) provide a review on studies showing that humans learn and remember information better when experienced multimodally as opposed to unimodally.

than participants in the unimodal speech-only condition. The difference between experimental groups was particularly pronounced when participants' recall of both the first *and* the last item was considered, yielding a significant effect of condition.

In view of the previously mentioned concerns regarding participants' potentially limited attention to the robot while solving the task, note that the information requested for recall was conveyed in the first utterance chunk of the instruction. This means, while the robot was describing the object that was next to be moved, participants were still paying full attention to the delivery of the instruction.

Generally, the results support the view that providing information via multiple modalities, especially by adding imagery through the use of gesture, may aid cognitive processes such as information processing and recall in human observers. These findings are in line with human gesture research which suggests that the use of gesture can be beneficial to human listeners when retrieving priorly communicated information (e.g., Kelly et al., 1999; Galati and Samuel, 2011). The fact that a similar effect was found to apply for robot gesture in the present study suggests that humans are able to naturally process and benefit from such non-verbal behavior even when it is conveyed by an artificial communicator.

Chapter 8

Empirical Evaluation - Study 2

Based on the experiences gained from the first exploratory study described in the previous chapter, the second experimental study aimed at a more focused investigation. Findings from Study 1 led to a number of questions with regard to the generalizability of the results. First, was the trend towards higher ratings regarding the *perception of the robot* in the incongruent multimodal condition compared to the congruent condition only a 'coincidence' because participants were not focusing on the robot? Second, was this also the reason for the comparatively good *task-related performance of participants* in the incongruent condition, despite the partly contradictory information conveyed in the two modalities?[1] Third, if the robot's use of co-verbal gesture already improved participants' *information uptake* with only little attention paid to the robot, would this effect be even more pronounced with their increased focus on the robot's non-verbal behavior?

In view of these questions emerging from the initial study, the second experiment described in this chapter was intended to complement the previous findings based on a modified experimental design and a more specific data analysis. In addition, it aimed at shedding light on how communicative non-verbal behaviors affect social perceptions of the robot and the HRI experience. Therefore, a central purpose of the present study was to investigate how robot gesture would affect anthropomorphic inferences about the humanoid robot. Specifically, data analysis regarding the *perception of the robot* focused on the attribution of typically human traits to the robot, evaluation of its likability, participants' perceived shared reality, and their future contact intentions after interacting with the robot. The methodology of Study 2 is described in Section 8.1; results are subsequently presented in Section 8.2.

---

[1]Notably, none of the participants of the first study made use of the table provided for the case that a participant did not fully understand the robot's instruction.

8.1 Method

In Study 1, it was often observed that participants immediately turned to the object being referred to by the robot within the first chunk of the utterance. In such cases the participant's attention typically shifted from the robot to the named object while the robot was still delivering the following chunk(s) of the instructional utterance. This behavior was reflected in the fact that participants of Study 1 often reported difficulties in assessing the robot's behavior after completing the task, since they had not consciously paid attention to the robot. As a consequence, the experimental design of the first study was modified so that the participants' attention would be directed towards the robot for a longer period of time during the interaction.

8.1.1 Experimental Design

The general set-up, scenario, and experimental conditions in Study 2 were similar to the design of Study 1. However, in order to increase the participants' attention towards the robot, each utterance was delivered by the robot in two parts. The first part referred to the object (e.g., *"Please take the thermos flask"*), which corresponds to the first chunk of a two-chunk or three-chunk utterance (cf. Section 7.1.1). The second part comprised the designated location and position of the item (e.g., *"...and place it on the right side of the upper cupboard."*), which corresponds to the second chunk of a two-chunk utterance or the second and third chunk of a three-chunk utterance. In the multimodal conditions, the gestures maintained their synchronization with the verbal chunks, thus gestural behavior was effectively paused whenever there was a break in the delivery of the utterance. **Figure 8.1** illustrates the modified experimental design.

Although potentially a less natural form of interaction, the primary motivation in splitting the utterances was to increase the participants' attention directed towards the robot. The second part of the utterance was only triggered once the participant had picked up the object from the box and had returned to stand in front of the robot, while directing their gaze at the robot in anticipation of the next instruction. This way, it was assured that participants observed the complete set of chunks delivered by the robot.

"Please take the thermos flask"   [robot pauses while participant retrieves object]   "and place it on the right side of the upper cupboard."

Figure 8.1: Modified experimental design with split utterances: the robot's utterance delivery is paused after the first chunk until the participant has retrieved the object.

8.1.2 Experimental Procedure

The experimental procedure in Study 2 was almost identical to Study 1, with the only difference being the modified delivery of utterance chunks. Moreover, in Study 1, participants' verbal interaction with the robot was observed to be minimal, typically limited to the word "next". It was consequently assumed that participants had interpreted the experimental instructions to imply that the robot only understood this one word, and thus they avoided any other form of verbal interaction. Therefore, to circumvent this perceived limitation in the interaction, participants in Study 2 were not required to verbally indicate when they were ready for the robot to proceed with the next piece of information; instead, they were asked to stand in front of the robot to receive the next piece of information. That is, whenever the participant resumed a standing position in front of the robot to signal readiness to proceed with the next instruction, the experimenter sitting at a control terminal triggered the robot's subsequent behavior. Finally, in contrast to Study 1 in which a more subtle and natural perception of the robot was achieved, in Study 2 the experimenter explicitly instructed the participants to dedicate their attention towards the robot while solving the given task.

8.1.3 Hypotheses

Based on the findings from Study 1 and in consideration of participants' increased focus on the robot, the following hypotheses were established for the utilization of split utterances in Study 2:

H4: Participants who receive multimodal instructions from the robot (either congruent or incongruent) will evaluate the robot more positively and anthropomorphize it more than those who receive unimodal information (i.e., speech-only). Moreover, similar to the findings from Study 1, a robot that occasionally performs non-matching gestures will be preferred over one that performs no gestures at all.

H5: With a greater focus on the robot's behavior, participants who are presented with incongruent multimodal instructions by the robot will perform worse at the task than those who are presented with unimodal or congruent multimodal information by the robot.

H6: Participants who receive multimodal instructions from the robot will take up more information about the objects handled during the interaction than those who are presented with unimodal information by the robot. Trends observed in Study 1 should be more pronounced in Study 2 due to the participants' increased attention to the robot's behavior.

8.1.4 Dependent Measures

As in Study 1, participants were asked to report their interaction experience with the robot and rate their perception of its behavior based on a post-experimental questionnaire. Video data was analyzed to evaluate participants' performance during the task. Data analysis for Study 2 focused on the same four main aspects as in the first study, namely *quality of presentation*, *perception of the robot*, *task-related performance of participants*, and *information uptake* (see Section 7.1.4).

However, with regard to participants' *perception of the robot*, a different set of measures was used for a more specific analysis based on social psychological research. Specifically, the degree of anthropomorphism attributed to the robot was assessed using measures from research on the (de-)humanization of social groups (Haslam et al., 2008). To illustrate, Haslam et al. have proposed two distinct senses of humanness at the trait level by differentiating between 'uniquely human' and 'human nature' traits. While 'uniquely human' traits imply higher cognition, civility, and refinement, traits indicating 'human nature' involve emotionality, warmth, desire, and openness. Since the human nature dimension is typically used to measure 'mechanistic dehumanization'[2], this measure was conversely employed

---

[2]According to Haslam et al. (2008), when people are denied human nature, they are implicitly or explicitly objectified or likened to machines rather than to animals or humans.

to assess the extent of the robot's perceived **human-likeness**. Haslam et al.'s list of human nature traits comprises ten characteristics which are presented in **Table 8.1** together with other measures used to evaluate participants' *perception of the robot* in Study 2. In particular, participants' perception of the robot's **likability** was assessed using three questionnaire items. Their degree of **shared reality** with the robot was evaluated based on three further items which tap perceptions of similarity and experienced psychological closeness to the robot (Echterhoff et al., 2009). The shared reality index also covers aspects of human-robot acceptance, as participants had to indicate how much they enjoyed the interaction with the robot. Finally, participants' **future contact intentions** with regard to the robot were measured using a single item. All indices and measures together with the respective questionnaire items and scales used to evaluate participants' *perception of the robot* in the second study are summarized in Table 8.1.

Dependent measures used for the evaluation of the *quality of presentation, task-related performance of participants,* and *information uptake* were identical to the measures applied in Study 1 (see Tables 7.2, 7.4, and 7.5 respectively).

Table 8.1: Dependent measures, respective questionnaire items and scales used to evaluate the *perception of the robot* in Study 2:

Index / Measure:	Items:	Scale:
Human-Likeness	curious, friendly, fun-loving, sociable, trusting, aggressive, distractible, impatient, jealous, nervous	1 = strongly disagree, 5 = strongly agree
Likability	polite, sympathetic, humble	1 = strongly disagree, 5 = strongly agree
Shared Reality	"How close do you feel to the robot?" "How pleasant was the interaction with the robot for you?" "How much fun did you have interacting with the robot?"	1 = not at all, 5 = very much
Future Contact Intentions	"Would you like to live with the robot?"	1 = not at all, 5 = very much

8.1.5 Participation

A total of 62 participants (32 female, 30 male) took part in the experiment, ranging in age from 20 to 61 years ($M = 30.90$ years, $SD = 9.82$). All participants were German native speakers recruited at Bielefeld University, Germany. Based on five-point Likert scale ratings, participants were identified as having negligible experience with robots ($M = 1.35$, $SD = 0.66$) and moderate skills regarding technology and computer use ($M = 3.74$, $SD = 0.97$). Participants were randomly assigned to one of the three experimental conditions that manipulated the robot's non-verbal behaviors, while maintaining gender- and age-balanced distributions.

8.2 Results

8.2.1 Quality of Presentation

Similar to the analyses conducted for the first study, the quality of presentation was assessed based on participants' ratings with regard to gesture and speech generated by the robot. As before, for the unimodal condition only gesture quantity was measured. Mean values and standard deviations are listed in **Table 8.2**.

Table 8.2: Mean values of dependent measures rating the *quality of presentation* in the three conditions of Study 2 (standard deviations in parentheses).

		Condition	
Measure	**Unimodal**	**Congr. Multimodal**	**Incongr. Multimodal**
Gesture Quantity	1.22 (0.65)	3.35 (0.67)	3.38 (0.59)
Gesture Speed		2.95 (0.39)	2.86 (0.36)
Gesture Fluidity		3.95 (1.10)	3.95 (0.81)
Speech-Gesture Content		4.10 (0.85)	2.48 (1.29)
Speech-Gesture Timing		4.05 (0.83)	4.05 (0.81)
Naturalness		3.14 (0.96)	3.67 (1.11)

With regard to *gesture quantity*, the absence of the robot's non-verbal behaviors resulted in very low mean ratings from participants in the unimodal condition ($M = 1.22$, $SD = 0.65$). In contrast, the overall mean value for the two gesture conditions was $M = 3.37$ ($SD = 0.62$), indicating that participants in the multimodal groups were generally satisfied with the gesture rate. When comparing the three experimental groups, an analysis of variance (ANOVA) indicated a significant effect of condition regarding the *gesture quantity* ratings ($F(2,56) = 71.18$, $p < .001$). Pairwise comparisons with Tukey's post-hoc test confirmed significantly higher ratings in both the congruent multimodal group when compared to the unimodal group, $p < .001$, and the incongruent multimodal group when compared to the unimodal group, $p < .001$. A comparison of the two multimodal conditions was not found to yield significant differences.

When combining the results of the two multimodal conditions, the remaining five attributes measuring presentation quality resulted in the following overall mean values (standard deviation in parentheses). Evaluations of *gesture speed* yielded $M = 2.90$ ($SD = 0.37$), which is consistent with the results from Study 1. That is, on average the generated gestures were considered to be of appropriate speed. *Gesture fluidity* was rated at $M = 3.95$ ($SD = 0.95$) across the two multimodal conditions with identical mean values in each condition. This implies that the robot's gestures were perceived as fairly smooth and fluid, which reflects a more positive mean rating in comparison to Study 1. Semantic matching of speech and gesture (*speech-gesture content*) was rated at $M = 3.27$ ($SD = 1.36$). This time, however, a comparison of the two multimodal groups using an independent-samples t-test showed a highly significant effect ($t(34,84) = 4.78$, $p < .001$) with higher ratings in the congruent compared to the incongruent multimodal condition. This difference suggests that in Study 2, participants in the incongruent multimodal group took more notice of the partly non-matching gestures, potentially due to their increased attention directed toward the robot. Temporal matching of speech and gesture (*speech-gesture timing*) was again perceived as fairly appropriate with $M = 4.05$ ($SD = 0.81$) and identical mean values in both gesture conditions. Finally, perceived *naturalness* of the robot's combined use of speech and gesture was rated slightly better than in Study 1, with an overall mean value of $M = 3.40$ ($SD = 1.06$) for the two gesture conditions. Interestingly, ratings of perceived naturalness were notably higher in the incongruent condition than in the congruent gesture condition, although this difference was not found to be significant.

In summary, as with Study 1, the results suggest that participants were generally satisfied with the presentation quality of the gestures performed by the robot. Furthermore, the modification of the study design appears to have affected ratings regarding the semantic matching of speech and gesture, indicating that participants in the third experimental group recognized the partial incongruity regarding the robot's utterances.

8.2.2 Perception of the Robot

First, reliability analyses (Cronbach's α) were conducted to assess the internal consistencies of the dependent measures where applicable. The proposed indices proved sufficiently reliable, given a Cronbach's α of .78 for the *human-likeness* index, $\alpha = .73$ for the *likability* index, and $\alpha = .78$ for the *shared reality* index. Consequently, participants' responses to the corresponding questionnaire items were averaged to form the three outlined indices. To test the effect of experimental condition on the dependent measures, analyses of variance (ANOVA) and Tukey's post-hoc tests were conducted with a 95% confidence interval for pairwise comparisons between condition means. Mean values and standard deviations are summarized in **Table 8.3** and are visualized in **Figure 8.2**.

Results show a significant effect of condition on all dependent measures. Specifically, they confirm that the manipulation of the robot's gestural behavior had a significant effect on participants' ratings of the *human-likeness* index which reflects their attribution of human nature traits to the robot ($F(2,58) = 4.63$, $p = .014$). It also had a significant effect on their assessment of the robot's *likability* ($F(2,59) = 3.65$, $p = .032$). Furthermore, analyses indicate that the manipulation of the robot's non-verbal behavior had a significant effect on participants' ratings of the *shared reality* measure ($F(2,59) = 4.06$, $p = .022$) as well as on their *future contact intentions* ($F(2,58) = 5.43$, $p = .007$).

Tukey post-hoc comparisons of the three groups indicate that participants in the incongruent multimodal condition ($M = 2.55$, $SD = 0.68$) rated the perceived *human-likeness* of the robot significantly higher than participants in the unimodal condition ($M = 1.98$, $SD = 0.58$), $p = .007$. That is, when the robot performed gestures that were to some extent incongruent with speech, participants anthropomorphized it significantly more than when it did not gesture at all.

Moreover, participants reported significantly greater perceived *likability* when interacting with the robot whose verbal utterances were accompanied by partly non-

Table 8.3: Mean values of the dependent measures reflecting participants' *perception of the robot* (standard deviations in parentheses).

	Condition		
Measure	Unimodal	Congr. Multimodal	Incongr. Multimodal
Human-Likeness	1.98 (0.58)	2.15 (0.58)	2.55 (0.68)
Likability	3.69 (0.97)	3.92 (0.81)	4.36 (0.59)
Shared Reality	3.23 (0.93)	3.75 (0.76)	3.92 (0.70)
Future Contact Intentions	2.63 (1.30)	2.95 (1.40)	3.90 (1.14)

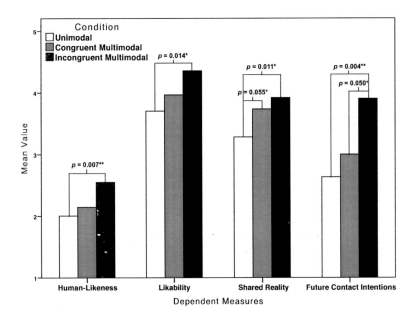

Figure 8.2: Bar chart visualizing the mean ratings and significant effects for the dependent variables measuring participants' *perception of the robot*; $^{+} = p < .10$, $^{*} = p < .05$, $^{**} = p < .01$.

matching gestures in the incongruent multimodal condition ($M = 4.36$, $SD = 0.59$) than when it was only using speech ($M = 3.69$, $SD = 0.97$), $p = .014$.

Participants also experienced greater *shared reality* with the robot when it used either congruent ($M = 3.75$, $SD = 0.76$) or incongruent ($M = 3.92$, $SD = 0.70$) multimodal behaviors than when it relied on unimodal communication only ($M = 3.23$, $SD = 0.93$); this effect was marginally significant for the comparison of unimodal versus congruent multimodal behavior, $p = .055$, and significant when comparing the unimodal with the incongruent multimodal condition, $p = .011$.

Finally, participants' assessment of *future contact intentions* with regard to the robot was also significantly higher in the condition with partially incongruent speech-accompanying gesture behavior ($M = 3.90$, $SD = 1.14$) than in the unimodal condition ($M = 2.63$, $SD = 1.30$), $p = .004$. Remarkably, average ratings of whether participants would like to live with the robot were much higher in the incongruent multimodal condition than in the congruent multimodal condition group ($M = 2.95$, $SD = 1.40$), just missing significance with $p = .050$.

Although comparisons between the unimodal and the congruent multimodal condition were not statistically significant at the 5 % level, they indicate a trend towards higher mean ratings for all dependent measures in the congruent multimodal condition. Similarly, comparisons between the congruent multimodal and the incongruent multimodal groups were not statistically significant at $p < 0.05$, however, the results throughout indicate a trend towards higher mean ratings in favor of the incongruent multimodal group. These observed trends with regard to participants' *perception of the robot* are in line with the results from Study 1. Furthermore, they support hypothesis H4 which predicted higher ratings on all dependent measures in the two multimodal groups when compared to the unimodal group.

8.2.3 Task-Related Performance of Participants

As with Study 1, participants' *task-related performance* was measured subjectively based on self-ratings using the questionnaire item displayed in Table 7.4 as well as objectively based on the task-related error rate. The objective measure was used to test the respective hypothesis (H5).

Results of participants' *subjective assessment* ratings are shown in **Table 8.4**. In line with the results from Study 1, participants generally rated their own competence as high, with mean values between 4.05 and 4.60 in all three groups. However,

Table 8.4: Mean values of the measure indicating participants' *subjective assessment* (standard deviations in parentheses).

| | Condition | | |
Measure	Unimodal	Congr. Multimodal	Incongr. Multimodal
Competence Self-Rating	4.60 (0.50)	4.67 (0.58)	4.05 (0.81)

a one-way ANOVA indicated a significant effect of experimental condition on participants' self-ratings regarding their task-related competence, $F(2,59) = 5.83$, $p = .005$. Pairwise comparisons with Tukey's post-hoc test further revealed that participants in the incongruent multimodal group ($M = 4.05$, $SD = 0.81$) rated their own performance significantly worse than participants both in the unimodal group ($M = 4.60$, $SD = 0.50$), $p = .021$, and in the congruent multimodal group ($M = 4.67$, $SD = 0.58$), $p = .008$.

Results of *objective assessment* ratings are summarized in **Table 8.5** and are illustrated in **Figure 8.3**. In contrast to the results from Study 1, the average error rate across all nine kitchen objects handled in the experiment was found to be highest for the incongruent multimodal condition with a total average error rate of 11.12 %. This comprises an error rate of 9.53 % with regard to misplaced objects and an additional mean error of 1.59 % with regard to objects that were placed on the adjacent table, indicating that the participant had failed to understand the robot's instruction. In comparison, average error rates were much lower in the other two conditions with a combined error rate of 2.78 % (2.22 % misplaced objects) in the unimodal condition and a combined error rate of 2.65 % (2.12 % misplaced objects) in the congruent multimodal condition. In fact, a Kruskal-Wallis test showed a significant effect of condition, $\chi^2(2) = 9.06$, $p = .011$. Mann-Whitney tests were conducted with a Bonferroni correction to follow up on this finding, yielding a significant difference both between the unimodal and the incongruent multimodal groups ($U = 127.50$, $p = .008$, $r = -.38$) and between the congruent and incongruent multimodal groups ($U = 132.00$, $p = .006$, $r = -.39$). That is, in accordance with hypothesis H5, participants who received partly incongruent instructions from the robot performed significantly worse than participants who received either unimodal or congruent multimodal instructions from the robot. These results indicate that, with an increased focus

Table 8.5: Average error rates per object and group used to measure participants' *objective assessment*; values in brackets refer to objects placed on the adjacent table; asterisks (*) indicate items which were described in utterances with non-matching gestures in the incongruent multimodal condition.

| Object | | Condition | |
	Unimodal	Congr. Multimodal	Incongr. Multimodal
Thermos Flask	0 %	4.8 %	9.5 %
Rect. Chopping Board*	10.0 %	0 %	9.5 %
Large Bowl*	0 %	0 %	42.9 %
Egg Cup*	0 %	0 %	4.8 % (14.3 %)
Vase	0 % (5.0 %)	0 %	0 %
Sieve*	0 %	0 %	0 %
Small Bowl*	0 %	0 %	0 %
Soup Ladle*	0 %	0 %	4.8 %
Round Chopping Board	10.0 %	14.3 % (4.8 %)	14.3 %
Average Error	2.22 % (0.56 %)	2.12 % (0.53 %)	9.53 % (1.59 %)

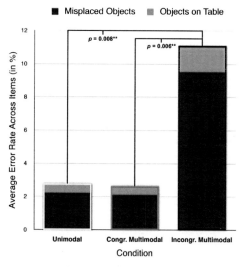

Figure 8.3: Bar chart visualizing participants' *objective assessment* based on the average error rate per group across all nine objects handled in the experimental task.

on the robot's behavior in Study 2, the partly non-matching gestures affected the participants' perception of the robot's instructions and, ultimately, had a negative impact on their performance.

Finally, as observed in Study 1, Spearman's correlation analysis showed a significant negative correlation between *objective* and *subjective* assessment measures ($r = -.41$, $p = .001$). That is, participants' self-ratings were generally in line with their objective assessments: the more mistakes participants made in the experimental task, the lower they rated their own competence in the questionnaire.

8.2.4 Information Uptake

Participants' *information uptake* was measured based on the same two questions as used in Study 1 (see Table 7.5), namely asking participants to recall the first and last object they had to move during the experimental task. These results were designated for testing hypothesis H6 which, in view of the participants' increased focus on the robot during interaction, predicted a similar trend as observed in Study 1, however, with a more pronounced effect.

Participants' average recall rates for each experimental group regarding the first and last object are summarized in **Table 8.6** and illustrated in **Figure 8.4**. Across all three measures (i.e., correct recall of the first object, the last object, and both the first and last object) recall rates in the incongruent multimodal group were lower than in the unimodal and congruent multimodal groups. This trend is particularly pronounced with regard to the recall of the first object and the recall of both the first and the last object. Although not yielding a statistically significant effect, this trend opposes the findings from Study 1 and contradicts the prediction of H6. Furthermore, differences between the recall rate of the unimodal and the congruent multimodal condition were less pronounced than in the first study, but still show slightly higher recall rates in the congruent multimodal condition. Remarkably, recall rates of the last item were quite similar across all experimental conditions, yielding 85 % recall in the unimodal group versus 85.7 % in the congruent multimodal and 81 % in the incongruent multimodal group.

Generally, results regarding participants' *information uptake* indicate that an increased focus on the robot during interaction did not promote the effect of higher recall rates in the two multimodal groups as observed in the first study. Hence, the hypothesized predictions of H6 were rejected.

Table 8.6: Participants' *information uptake* per group based on the average recall rate regarding the first and last object moved during the experimental task.

	Condition		
Measure	Unimodal	Congr. Multimodal	Incongr. Multimodal
First Object Recall	50 %	61.9 %	33.3 %
Last Object Recall	85 %	85.7 %	81 %
First & Last Object Recall	50 %	57.1 %	28.6 %

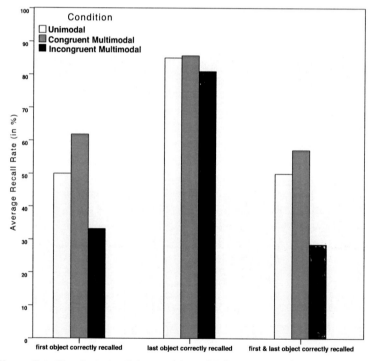

Figure 8.4: Bar chart visualizing participants' *information uptake* per group based on the average recall rate regarding the first and last object moved during the experiment.

8.3 Summary and Discussion

This chapter overviewed a second experimental study designed to complement and validate the findings from Study 1. Slight modifications of the experimental design and procedure were described in **Section 8.1**. Crucially, the study was modified to increase the participants' attention directed towards the robot during interaction by splitting the robot's utterance delivery into two parts. That is, the first utterance chunk was followed by a pause until the participant had retrieved the object being referred to. In addition, participants were explicitly instructed to direct their attention to the robot while solving the task. In this way, it was ensured that participants observed the complete set of chunks delivered by the robot. Remaining aspects of the experimental method, especially with regard to the design (i.e., general set-up, scenario, and experimental conditions) and the procedure, were similar to those of Study 1. Hypotheses proposed in Section 8.1.3 predicted a positive effect of robot gesture on dependent variables measuring participants' *perception of the robot* and *information uptake*. In addition, a negative effect of incongruent gesture was hypothesized with regard to participants' *task-related performance*. Detailed empirical results were reported in **Section 8.2**. The main findings with regard to the proposed hypotheses are interpreted and discussed in the following.

Perception of the Robot

In a more focused approach to investigating the effect that non-verbal behavior may have on participants' perception and evaluation of the robot, responses to 17 questionnaire items, distributed over three indices and a single item measure, were analyzed. These were used to evaluate participants' attribution of *human-likeness* to the robot, its perceived *likability*, as well as *shared reality* and *future contact intentions* with regard to the robot.

The results support the respective hypothesis H4 by showing that the robot's gestural behavior tends to result in a more positive subsequent evaluation of all dependent measures by the human participants. Intriguingly though, this observation was only statistically significant at the 5 % level when the incongruent multimodal condition was compared to the unimodal condition. That is, when the robot performed partly non-matching gestures, it was perceived and rated more positively than when it only used speech or when it performed congruent multimodal behavior (see RQ6). Specifically, with regard to the robot, this means that partly incongruent multimodal behavior resulted in greater perceived human-likeness, likability, shared reality, and future contact intentions. Despite the modification of the experimental design in Study 2, which increased participants' attention to the robot, these findings show similar trends to those of Study 1, suggesting that they were unlikely to be a coincidence.

Indeed, the results actually exceed the hypothetical expectations: not only do they indicate that a robot with occasional incorrect gestures is more favorable than a non-gesturing robot; they surprisingly suggest that human interaction partners even favor such partly incongruent multimodal behavior over completely matching multimodal behavior. At first, this finding appears counterintuitive – how can it therefore be interpreted?

The present analyses particularly focused on participants' attribution of typically human traits to the robot and resulting anthropomorphic inferences. The results may be better understood if the robot's partly incongruent co-verbal gestures are not just considered as non-matching utterances, but as unpredictable behavior. From this perspective, the present findings are actually in line with previous research on anthropomorphism and social robots suggesting that implementing some form of unpredictability in a robot's behavior can create an illusion of the robot being 'alive' (Duffy, 2003). Thus, participants in this group may have attributed intentions to the robot based on its unpredictable behavior, e.g., by assuming that it deliberately tried to confuse them; indeed, several participants in the incongruent multimodal group approached the experimenter after the interaction, reporting that the robot was "cheeky" or was "trying to fool" them.

An alternative interpretation of the results is that participants perceived the robot's incongruent gestures as errors or 'imperfections' which made the robot appear more human-like and less machine-like, and as a result, generally more likable (see RQ7). These interpretations of the results suggest that a certain level of unpredictability or 'imperfection' in a humanoid robot, as given in the incongruent gesture condition, can actually lead to a greater attribution of human traits to the robot and a more positive HRI experience. Although this observation certainly depends on the given context and task, e.g., whether or not the correctness and reliability of the robot's behavior are vital, it could potentially lead to a paradigm shift in the design of the 'perfect' social robot or artificial companion. Therefore, a clear need exists to further elucidate this finding in future HRI research.

Task-Related Performance of Participants

Based on the number of objects that were not correctly placed at their designated locations, the analysis of participants' task-related performance revealed a significant effect of condition in favor of hypothesis H5. That is, with an increased focus on the robot's behavior, participants who received partly incongruent instructions from the robot performed significantly worse in the task than participants who were presented with either unimodal or congruent multimodal instructions.

Remarkably, the greatest error rate of a single object, namely the "large bowl" (42.9 %), was found in the incongruent multimodal group. To find out a possible reason for this, it is useful to look at how the robot described the storage location of the object. As such, the robot verbally referred to the upper cupboard, but pointed to the top of the lower cupboard. In German, the words for 'upper' (*oberen*) and 'lower' (*unteren*) sound remotely similar, which may have led to a sensory fusion of the verbal and visual input so that the incongruity remained unnoticed by participants who followed the gestural instruction. Moreover, in contrast to Study 1, participants in this study also made use of the table which was provided for those instances in which a participant failed to understand the robot's instruction.

These results suggest that in Study 2, the partly non-matching gestures affected the participants' perception of the robot's instructions and had a negative impact on their performance. This interpretation is further supported by the fact that participants in the incongruent multimodal group rated their own competence at solving the task significantly lower than participants in the other two groups. The observed correlation between subjective and objective assessment measures thus indicates good self-assessment on the part of the participants.

In view of this finding, it appears even more surprising that the mean ratings of the dependent variables measuring participants' perception of the robot were highest in the incongruent condition. That is, although the robot's behavior negatively affected the participants' task-related performance, they still rated the robot as being more likable, reported greater shared reality with the robot, and expressed a greater desire to live with it than participants in the other groups. These findings therefore emphasize the positive impact of the incongruent gesture condition on participants' evaluation of the robot and should be systematically investigated in future studies.

Information Uptake

The last hypothesis (H6) of Study 2 predicted a similar trend as observed in Study 1, i.e., higher recall rates from participants in the two multimodal conditions regarding the first and last object handled during the experimental task. In view of the participants' increased focus on the robot during interaction, the effect was expected to be even more pronounced in Study 2.

However, recall rates from participants of the incongruent multimodal group were lower than for the unimodal and congruent multimodal groups across all three measures. This trend was found to be particularly pronounced with regard to the recall of the first object and the recall of both the first and the last object, although the difference did not reach significance. Thus, the results are not in line with findings from the first study and contradict the hypothesized predictions of H6. How can this observation be accounted for?

A possible interpretation is that, given the participants' increased focus on the robot's incongruent multimodal behavior and the accordingly higher task-related error rate, the task was cognitively more demanding for participants in this group. Therefore, the contradictory information delivered by the robot may have especially affected their memory of the first object – but not so much of the second object – to be moved during the interaction: by the time participants filled out the respective questionnaire item, more time had elapsed since the robot's delivery of the first instruction compared to the instruction regarding the last object. In addition, a set of incongruent instructions that followed the first object had to be processed during interaction, whereas no further instructions were received after the last object.

Furthermore, average recall for the unimodal condition was found to be only slightly lower than for the congruent multimodal condition and higher than for the incongruent multimodal condition. It is possible that, in contrast to the findings of Study 1, the delivery of split utterances in this study may have caused the increased recall rate in the speech-only condition, as participants generally spent more time moving each object. To illustrate, in the first study participants picked up each object and immediately stored it in the cupboard; in Study 2, participants had to pause in front of the robot while holding the object and awaiting the following instruction. This suggests that the advantage of multimodal versus unimodal representation decreases when participants are exposed to the objects for a longer period of time.

In summary, results regarding participants' *information uptake* show that increasing the focus on the robot's gestural behavior during interaction does not result in higher recall rates in the two multimodal groups compared to the unimodal condition. Instead, the results suggest that gestures may have a more supportive effect on information uptake in a more natural interaction scenario as provided by the robot's 'one-shot' utterance delivery of the first study.

"We can only see a short distance ahead, but
we can see plenty there that needs to be done."
Alan Turing

Chapter 9

Conclusion

The work described in this thesis focused on the challenges of generating and evaluating communicative gesture for a humanoid robot. It was motivated from both a technically and a psychologically inspired perspective: firstly, it was explained how only few scientific approaches had so far addressed the design and flexible generation of finely synchronized speech and gesture for robots; secondly, an examination of prior research in social human-robot interaction suggested a lack of systematic evaluation of the impact of such non-verbal behaviors on people's perception of the robot and on their general HRI experience.

This thesis therefore set out to contribute to and complement existing work by setting the precedent for a more rigorous and systematic study of conceptual motorics research. The adopted interdisciplinary approach spanned several sub-fields including psycholinguistics, computer science, neurobiology, engineering, and social psychology. The major contributions and findings of the present work are summarized in Section 9.1. Concluding this thesis, the scope for future research direction is outlined and desirable extensions to the contributed work are discussed in Section 9.2.

9.1 Summary of Contributions and Findings

Two major objectives addressing both technically and psychologically inspired research questions were outlined in this thesis. Working towards these objectives has resulted in the following contributions and findings in each respective field.

9.1.1 Technical Contributions and Implications

With regard to the technical implementation realized within the scope of the present work, the contributions are twofold:

199

9. CONCLUSION

1) **Speech-gesture generation framework for humanoid robots.** Building
on one of the most sophisticated virtual agent frameworks, this work provides
a novel multimodal action generation framework that is specifically tailored to
the requirements of speech and gesture synthesis for a humanoid robot. To
date, most gesture generation models in social robotics have only featured a
predefined repertoire of motor actions; the implemented system advances the
field by contributing a more flexible framework for the real-time production
of speech-accompanying robot gesture which is not restricted to a set of pre-
scripted motor primitives. Although the presented solution was designed and
realized for a specific robotic platform, it represents a proof of concept by
demonstrating the feasibility of the chosen approach. As a result, the work of
this thesis paves the way for similar approaches to employing a virtual agent
framework for behavior realization in arbitrary physical humanoid robots.

2) **Novel multimodal scheduler for closed-loop control.** All pre-existing
approaches to the generation of multimodal robot behaviors are based on
unidirectional synchronization mechanisms, in which gesture timing typically
adapts to the timing of speech, and open-loop control. That is, they do
not consider any sensory feedback to re-adjust the pre-planned behaviors at
run-time should adaptation be necessary, potentially resulting in asynchrony.
Besides the limited flexibility of those systems, previous approaches have
therefore failed to accurately model and account for the human ability to
mutually adapt speech and gesture to one another during the utterance process
(see de Ruiter, 1998, and Kendon, 2004, for empirical evidence). Representing a
technical novelty in social robotics research, the present system contributes the
first closed-loop approach to speech-gesture generation for humanoid robots. It
integrates two major features: first, the *planning* of multimodal behaviors was
optimized based on an empirically validated forward model which provides a
more accurate estimation of the robot's gesture preparation time; second, the
execution of behaviors was improved based on a reactive feedback mechanism
which enables the adaptation of running speech to generated gesture not only
at inter-chunk, but also at intra-chunk level, i.e., within a single multimodal
utterance chunk. The unprecedented quality of synchronized robot gesture
and speech as achieved by the proposed multimodal scheduler thus advances
the state of the art by providing a more flexible and natural way to realize
multimodal behavior for robots and other artificial communicators.

9.1.2   Empirical Findings and Implications

Based on the implemented technical framework, the psychologically motivated part of the present work set out to exploit the achieved flexibility in robot gesture generation for controlled experimental HRI studies. Involving a total of 122 participants in what to date represents two of the most comprehensive gesture-based interaction studies in social robotics, the empirical findings presented in this thesis substantially complement existing work in the field. Main findings and implications of each of the two studies are summarized in the following.

Conclusion and Main Findings of Study 1

To shed light on the impact of communicative robot gesture on human experience in HRI, a first exploratory study was conducted, inviting participants to interact with the robot in a joint task scenario. The robot's communicative behavior was manipulated in three experimental conditions to either include only speech without gesture, congruent gesture and speech, or partly incongruent gesture and speech. The results provide insights into how humans perceive and interpret the robot's utterances in relation to different communication modalities. Crucially, the use of gesture in addition to speech was found to enhance people's performance in the experimental task as well as their ability to take up and later recall information provided by the robot.

Besides emphasizing the task-related helpfulness and cognitive support of representational gesture in communication beyond human-human interaction, the results revealed a positive effect of multimodal behavior on human evaluation of the robot. In particular, they suggest that the inclusion of social cues in the form of co-verbal gesture casts the robot in a more positive light than when it is limited to a single communication modality, namely speech. Surprisingly, the robot was found to be evaluated as even more lively, active, engaged, friendly, communicative, and fun-loving when performing partly non-matching rather than fully congruent gestures. These findings indicate that a robot's non-verbal communicative behavior can trigger a more positive response within humans when it is not 'flawless'.

Representing a novel observation with regard to gesture-based HRI, these findings thus contribute to an advancement in social robotics and point out the direction for future HRI research dedicated to the design of acceptable artificial communicators.

Conclusion and Main Findings of Study 2

The second study complemented and partly supported the findings resulting from the initial study. Based on a modified experimental design to increase participants' attention to the robot and by applying a wide range of dependent measures, the experiment investigated how communicative gesture affects people's social perceptions and anthropomorphic inferences with regard to the robot.

The results revealed that with an increased focus on the robot's non-verbal behavior, incongruent robot gestures negatively affected participants' task-related performance and, potentially due to the cognitively more demanding task, their ability to memorize information presented to them. Importantly though, the main finding of Study 1 was replicated: when the robot performed hand gestures that were partly incongruent with accompanying speech, it was again rated more positively than when it only used speech or even when it performed congruent multimodal behavior. That is, although participants in this experimental group made more mistakes during the HRI task, the robot was anthropomorphized more and rated as more likable when it displayed incongruent multimodal behavior than when it performed congruent or no gesture. Furthermore, participants who were presented with partly non-matching gesture behavior felt closer to the robot and reported a significantly greater desire to live with it than participants in the other two experimental groups.

By reaffirming the positive impact of the incongruent gesture condition on participants' evaluation of the robot, these novel findings contribute to an advancement in HRI by encouraging a paradigm shift regarding the design of the 'perfect' social robot or artificial companion: not only may a certain degree of unpredictability be desirable in designing social agents (cf. Duffy, 2003); in addition, integrating occasional 'errors' into the agent's behavioral routine may in fact increase its perceived human-likeness and likability.

The pursued theory-driven approach was characterized by the application of social psychological theories of (de-)humanization to HRI (Haslam et al., 2008). By adapting these measures of anthropomorphism from research on human nature traits, the findings of this study complement existing work on the issue of measurement of anthropomorphism in social robotics. Thus, they also contribute to a deeper understanding of determinants and consequences of anthropomorphism.

General Implications of the Studies

In summary, the findings of the two experimental studies provide new insights into human perception and understanding of communicative gesture in robotic agents. They highlight the importance of displaying such non-verbal behaviors in social robots as significant factors that facilitate smooth and pleasant HRI. Since this robot prototype lacks visible facial features that could potentially enrich the interaction with human users, e.g., by conveying emotional states of the system, this further emphasizes the necessity to rely on additional communication channels such as gestural behaviors. Finally, by revealing the positive effect of the non-matching gesture condition, the studies provide a set of new empirical findings about the effects that co-verbal robot gesture may have on human interaction partners. In this way, the presented studies fundamentally contribute to the field of social robotics and pave the way towards novel approaches in designing and building better artificial communicators.

9.2 Outlook

The contributions of this thesis have unveiled a plenitude of potential research avenues to follow up on. Future work should extend the scope of this research both in technical and empirical aspects as elucidated in the following.

9.2.1 Desirable Technical Extensions

In view of the limited time frame that was provided for the realization of the present research, several potential extensions of the technical framework were outlined throughout the thesis and are summarized as follows.

- **Extending the generation pipeline.** As highlighted at the beginning of this thesis (Section 1.4), the present work focused on the production of multimodal robot behavior at the realization level of the behavior generation pipeline (see Figure 1.3). The technical contributions of this work thereby laid the cornerstone for an action generation framework that combines the implemented lower level functionalities with a high level 'cognitive architecture'. The realized approach provides on-line scheduling and generation of multimodal behavior that is not limited to a predefined action repertoire. Future work should further exploit this flexibility by extending the framework with content and behavior planning modules to allow for autonomous robot behavior based on conceptual motorics.

- **Improvement of forward model.** Section 6.2.1 provided an overview of possible approaches for the implementation of a forward model that can be employed to achieve more accurate predictions of gesture execution time required by the preparation phase. Three approaches, namely WBM-based trajectory simulation, Fitts' Law, and simple time estimation, were realized and evaluated. So far, none of these forward models provides a means for the robot to 'learn' the timing of its motor behavior or to improve future predictions based on past motor experiences as it is typical of biological systems. A desirable extension of the improved multimodal scheduler would thus incorporate machine learning algorithms, e.g., based on neural networks, to realize a more accurate forward model which would also be more plausible from a neurobiological point of view.

- **Transfer to other robotic platforms.** One of the main objectives of the present work was to provide a proof of concept to demonstrate that the utilization of a virtual agent framework represents a feasible approach to multimodal behavior realization in physical humanoid robots. In view of this objective, the proposed system was specifically designed and implemented to account for the demand of future generalizability. Accordingly, a future aim is to transfer the developed system to other arbitrary platforms, either robotic or virtual, with humanoid embodiment to further validate the presented approach.

9.2.2 Future Avenues of Empirical Research

The studies conducted as part of the present work revealed some interesting and unexpected findings which should be systematically investigated in future experiments. Potential scope for future research direction includes but is not limited to the following aspects.

- **Analysis of video data from the studies.** The conducted experiments yielded a huge amount of video data documenting the interaction between each participant and the robot. Given a total of 122 participants, the annotation and analysis of the complete video corpus would have exceeded the scope of the present work. Nevertheless, the collected video data represents a promising source of experimental information and should thus be exploited for further investigations. These should concentrate on a targeted analysis of participants' interaction experience, for example, to examine whether the use of gesture affected participants' attention or their interactive behavior toward the robot.

- **Replicability of main findings and investigation of long-term effects.**
  Future research should investigate the generalizability of the present findings regarding anthropomorphic inferences and incongruent modalities with other robotic and virtual platforms, ideally with different levels of embodiment (e.g., humanoid versus non-humanoid). For example, it may be possible that participants in the studies rated incongruent speech and gesture positively when presented by this specific robot, but may disapprove of such behavior when displayed by a robot with less anthropomorphic design. Importantly, studies replicating this work should try to shed light on the role of the given HRI task and context, as well as the extent to which incongruent behaviors result in similar effects. In this regard, future work should also investigate the validity of these findings in long-term interaction studies in order to examine if this novel effect may 'wear off' after some time or, in contrast, whether it may even keep the human-robot relationship more appealing.

- **Isolated analysis of different aspects of non-verbal behavior.** In the studies presented, the robot's gaze behavior in the multimodal conditions was modeled in a simplistic way. This design choice was made on purpose to direct the participants' attention rather to the hand and arm movements performed by the robot. As a consequence, however, the robot's gazing behavior did not appear fully natural during the interaction, as the robot did not follow the human interaction partner with its gaze. In future studies, it will be desirable to examine the impact of gaze behavior displayed by the robot in an isolated experimental set-up without hand and arm gesture. In this way, it can be investigated to what extent anthropomorphic inferences, likability, shared reality, and future contact intentions are determined by the robot's arm gesture versus by gaze alone. Another characteristic of the present studies was the use of self-sufficient speech to keep the verbal utterances consistent and the task solvable also in the unimodal condition. As emphasized by Hostetter (2011), however, representational gestures have the greatest impact on human listeners when they convey non-redundant information that is not contained in speech but is crucial for comprehension. For example, the utterance "put it there" is ambiguous without a spatial reference, e.g., as provided by a deictic gesture. In contrast, in the present studies the robot's gestures conveyed additional illustrative information which was not indispensable to solving the task. Therefore, to increase the measurable effect of robot gesture in a more focused study, the robot's non-verbal behaviors should be designed to contain complementary information necessary for successful interaction but which is not conveyed via speech.

In conclusion, the realized technical framework provides a suitable testbed for studying the effects of different gestural patterns in a highly controllable social interaction. By employing the robot as a tool to systematically investigate human perception of gesture for future research, the present work will not only advance the field of social robotics but also that of human gesture research. In view of the wide scope for future work, it is hoped that these research avenues will be tackled to further consolidate the contributions of the present work.

References

Abend, W., Bizzi, E., and Morasso, P. (1982). Human arm trajectory formation. *Brain*, 105:331–348.

Alibali, M. W. (2005). Gesture in spatial cognition: Expressing, communicating, and thinking about spatial information. *Spatial Cognition and Computation*, 5:307–331.

Alibali, M. W., Bassok, M., Solomon, K. O., Syc, S. E., and Goldin-Meadow, S. (1999). Illuminating mental representations through speech and gesture. *Psychological Science*, 10:327–333.

Alibali, M. W., Kita, S., and Young, A. J. (2000). Gesture and the process of speech production: We think, therefore we gesture. *Language and Cognitive Processes*, 15(6):593–613.

Bavelas, J. B. (1994). Gestures as Part of Speech: Methodological Implications. *Research on Language and Social Interaction*, 27(3):201–221.

Bavelas, J. B., Chovil, N., Lawrie, D. A., and Wade, A. (1992). Interactive Gestures. *Discourse Processes*, 15:469–489.

Bavelas, J. B., Gerwing, J., Sutton, C., and Prevost, D. (2008). Gesturing on the telephone: Independent effects of dialogue and visibility. *Journal of Memory and Language*, 58:495–520.

Bavelas, J. B., Kenwood, C., Johnson, T., and Phillips, B. (2002). An experimental study of when and how speakers use gestures to communicate. *Gesture*, 2(1):1–18.

Beattie, G. (2003). *Visible Thought: The New Psychology Of Body Language*. Routledge, London.

Beattie, G. and Shovelton, H. (2001). An experimental investigation of the role of different types of iconic gesture in communication: A semantic feature approach. *Gesture*, 1(2):129–149.

Beattie, G. and Shovelton, H. (2002). What properties of talk are associated with the generation of spontaneous iconic hand gestures? *British Journal of Social Psychology*, 41:403–417.

Bennewitz, M., Faber, F., Joho, D., and Behnke, S. (2007). Fritz – A Humanoid Communication Robot. In *Proceedings of the 16th IEEE International Symposium on Robot and Human Interactive Communication*.

Bergmann, K. (2011). *The Production of Co-Speech Iconic Gestures: Empirical Study and Computational Simulation with Virtual Agents*. PhD Dissertation, Bielefeld University.

Bergmann, K. and Kopp, S. (2009). Increasing the Expressiveness of Virtual Agents – Autonomous Generation of Speech and Gesture for Spatial Description Tasks. In Decker, K., Sichman, J., C., S., and C., C., editors, *Proceedings of the 8th International Conference on Autonomous Agents and Multiagent Systems*, pages 361–368.

Bergmann, K., Kopp, S., and Eyssel, F. (2010). Individualized Gesturing Outperforms Average Gesturing – Evaluating Gesture Production in Virtual Humans. In *Proceedings of the 10th Conference on Intelligent Virtual Agents*, pages 104–117. Springer.

Bernstein, N. (1967). *The Co-ordination and Regulation of Movements*. Pergamon Press, Oxford, UK.

Billard, A. (2001). Learning motor skills by imitation: A biologically inspired robotic model. *Cybernetics and Systems*, 32:155–193.

Billard, A., Calinon, S., Dillmann, R., and Schaal, S. (2008). Robot Programming by Demonstration. In Siciliano, B. and Khatib, O., editors, *Handbook of Robotics*, pages 1371–1394. Springer, Secaucus, NJ, USA.

Black, A. W. and Taylor, P. A. (1997). The Festival Speech Synthesis System: System documentation. Technical Report HCRC/TR-83, Human Communciation Research Centre, University of Edinburgh, Scotland, UK.

Breazeal, C. (2002). *Designing Sociable Robots*. AAAI Press.

Breazeal, C. (2003). Toward Sociable Robots. *Robotics and Autonomous Systems*, 42(3-4):167–175.

Bremner, P., Pipe, A., Melhuish, C., Fraser, M., and Subramanian, S. (2009). Conversational Gestures in Human-Robot Interaction. In *Proceedings of the 2009 IEEE International Conference on Systems, Man and Cybernetics*, pages 1645–1649.

Buisine, S., Abrilian, S., and Martin, J.-C. (2004). Evaluation of Multimodal Behaviour of Embodied Agents - Cooperation between Speech and Gestures. *From Brows to Trust*, 7:217–238.

Buisine, S., Abrilian, S., Niewiadomski, R., Martin, J.-C., Devillers, L., and Pelachaud, C. (2006). Perception of Blended Emotions: From Video Corpus to Expressive Agent. In Gratch, J., Young, M., Aylett, R., Ballin, D., and Olivier, P., editors, *Proceedings of the 6th International Conference on Intelligent Virtual Agents*, volume 4133 of *Lecture Notes in Computer Science*, pages 93–106. Springer.

Calinon, S. and Billard, A. (2007). Learning of Gestures by Imitation in a Humanoid Robot. In Dautenhahn, K. and Nehaniv, C., editors, *Imitation and Social Learning in Robots, Humans and Animals: Behavioural, Social and Communicative Dimensions*, pages 153–177. Cambridge University Press.

Cassell, J. (1996). Believable Communicating Agents - SIGGRAPH 1996 Course Notes.

Cassell, J. (1998). A Framework for Gesture Generation and Interpretation. In Cipolla, R. and Pentland, A., editors, *Computer Vision in Human-Machine Interaction*, pages 191–215, New York. Cambridge University Press.

Cassell, J., Bickmore, T., Campbell, L., Vilhjálmsson, H., and Yan, H. (2000). Human Conversation as a System Framework: Desigining Embodied Conversational Agents. In *Embodied Conversational Agents*, pages 29–63. MIT Press: Cambridge, MA.

Cassell, J., McNeill, D., and McCullough, K.-E. (1998). Speech-gesture mismatches: Evidence for one underlying representation of linguistic and nonlinguistic information. *Pragmatics & Cognition*, 6(2):1–34.

Cassell, J., Pelachaud, C., Badler, N., Steedman, M., Achorn, B., Becket, T., Douville, B., Prevost, S., and Stone, M. (1994). Animated Conversation: Rule-based Generation of Facial Expression, Gesture & Spoken Intonation for Multiple Conversational Agents. In *Proceedings of ACM SIGGRAPH 1994*, pages 413–420. ACM.

Cassell, J. and Thorisson, K. R. (1999). The Power of a Nod and a Glance: Envelope vs. Emotional Feedback in Animated Conversational Agents. *Applied Artificial Intelligence*, 13(4-5):519–538.

Cassell, J., Vilhjálmsson, H., and Bickmore, T. (2001). BEAT: the Behavior Expression Animation Toolkit. In *Proceedings of ACM SIGGRAPH 2001*.

Church, R. and Goldin-Meadow, S. (1986). The mismatch between gesture and speech as an index of transitional knowledge. *Cognition*, 23:43–71.

Clark, H. H. (1996). *Using Language*. Cambridge University Press, Cambridge.

Cohen, A. A. (1977). The Communicative Functions of Hand Illustrators. *Journal of Communication*, 27(4):54–63.

Cohen, A. A. and Harrison, R. P. (1973). Intentionality in the use of hand illustrators in face-to-face communication situations. *Journal of Personality and Social Psychology*, 28(2):276 – 279.

Condon, W. (1976). An analysis of behavioral organization. *Sign Language Studies*, 5(13):285–318.

Crystal, D. and Davy, D. (1969). *Investigating English Style*. Bloomington, Indiana University Press.

210

de Ruiter, J. P. (1998). *Gesture and Speech Production. PhD Thesis.* MPI Series in Psycholinguistics, University of Nijmegen.

de Ruiter, J. P. (2000). The Production of Gesture and Speech. In McNeill, D., editor, *Language and Gesture*, pages 284–311. Cambridge University Press, Cambridge, UK.

DeCarolis, B., Pelachaud, C. Poggi, I., and Steedman, M. (2004). APML, a markup language for believable behavior generation. In Prendinger, H. and Ishizuka, M., editors, *Life-like Characters. Tools, Affective Functions and Applications*, pages 65–85. Springer.

Desmurget, M. and Grafton, S. (2000). Forward modeling allows feedback control for fast reaching movements. *Trends in Cognitive Sciences*, 4(11):423–431.

Duffy, B. R. (2003). Anthropomorphism and the Social Robot. *Robotics and Autonomous Systems*, 42(3-4):177–190.

Echterhoff, G., Higgins, E., and Levine, J. (2009). Shared Reality: Experiencing Commonality with Others' Inner States about the World. *Perspectives on Psychological Science*, 4:496–521.

Efron, D. (1941). *Gesture and Environment.* King's Crown Press, New York.

Efron, D. (1972). *Gesture, Race, and Culture.* De Gruyter Mouton, Den Haag.

Ekman, P. and Friesen, W. V. (1969). The repertoire of nonverbal behavior: Categories, Origins, Usage, and Coding. *Semiotica*, 1:49–98.

Engle, R. A. (2000). *Toward a Theory of Multimodal Communication: Combining Speech, Gestures, Diagrams and Demonstrations in Instructional Explanations.* PhD Dissertation, School of Education, Stanford University.

Epley, N., Waytz, A., and Cacioppo, J. (2007). On Seeing Human: A Three-factor Theory of Anthropomorphism. *Psychological Review*, 114(4):864–886.

Feldman, A. G. (1986). Once more on the equilibrium-point hypothesis (lamda model) for motor control. *Journal of Motor Behavior*, 18:117–54.

Feyereisen, P. (2006). Further investigation on the mnemonic effect of gestures: Their meaning matters. *European Journal of Cognitive Psychology*, 18(2):185–205.

Feyereisen, P., Van de Wiele, M., and Dubois, F. (1988). The meaning of gestures: What can be understood without speech? *Cahiers de Psychologie Cognitive*, 8:3–25.

Fitts, P. M. (1954). The information capacity of the human motor system in controlling the amplitude of movement. *Journal of Experimental Psychology*, 47:381–391.

Flanagan, J. and Rao, A. (1995). Trajectory adaptation to a nonlinear visuomotor transformation: Evidence of motion planning in visually perceived space. *Journal of Neurophysiology*, 74:2174–2178.

Flash, T. and Hogan, N. (1985). The coordination of arm movements: an experimentally confirmed mathematical model. *Journal of Neuroscience*, 5:1688–1703.

Fong, T., Nourbakhsh, I. R., and Dautenhahn, K. (2003). A Survey of Socially Interactive Robots. *Robotics and Autonomous Systems*, 42(3-4):143–166.

Freedman, N. (1977). Hands, Words and Mind: On the Structuralization of Body Movements during Discourse and the Capacity for Verbal Representation. In Freedman, N. and Grand, S., editors, *Communicative Structures and Psychic Structures: A Psychoanalytic Interpretation of Communication*, pages 109–132. Plenum, New York.

Frick-Horbury, D. and Guttentag, R. E. (1998). The Effects of Restricting Hand Gesture Production on Lexical Retrieval and Free Recall. *American Journal of Psychology*, 111(1):43–62.

Galati, A. and Samuel, A. G. (2011). The role of speech-gesture congruency and delay in remembering action events. *Language and Cognitive Processes*, 26(3):406–436.

Ghez, C. and Krakauer, J. (2000). The Organization of Movement. In Kandel, E., Schwartz, J., and Jessell, T., editors, *Principles of Neural Science. Fourth Edition*, pages 653–673. McGraw-Hill Medical.

Gienger, M., Bolder, B., Dunn, M., Sugiura, H., Janssen, H., and Goerick, C. (2007). Predictive Behavior Generation – A Sensor-Based Walking and Reaching Architecture for Humanoid Robots. In Berns, K. and Luksch, T., editors, *Autonome Mobile Systeme 2007*, Informatik Aktuell, pages 275–281. Springer, Berlin, Heidelberg.

Gienger, M., Janssen, H., and Goerick, C. (2006). Exploiting Task Intervals for Whole Body Robot Control. In *Proceedings of the IEEE/RSJ International Conference on Intelligent Robots and Systems*, pages 2484–2490.

Gienger, M., Janßen, H., and Goerick, S. (2005). Task-Oriented Whole Body Motion for Humanoid Robots. In *Proceedings of the IEEE-RAS International Conference on Humanoid Robots*, Tsukuba, Japan.

Gienger, M., Toussaint, M., and Goerick, C. (2010). Whole-body Motion Planning Building Blocks for Intelligent Systems. In Harada, K., Yoshida, E., and Yokoi, K., editors, *Motion Planning for Humanoid Robots*, pages 67–98. Springer, 1st edition.

Gleicher, M. (1998). Retargeting Motion to New Characters. In *Proceedings of ACM SIGGRAPH 1998*, Annual Conference Series, pages 33–42.

Goldin-Meadow, S. (1999). The Role of Gesture in Communication and Thinking. *Trends in Cognitive Science*, 3:419–429.

Goldin-Meadow, S. (2003). *Hearing Gesture: How our Hands Help us Think*. Harvard University Press, Cambridge, MA.

Goldin-Meadow, S., Wein, D., and Chang, C. (1992). Assessing Knowledge Through Gesture: Using Children's Hands to Read Their Minds. *Cognition and Instruction*, 9(3):201–219.

Gordon, J., Ghilardi, M., and Ghez, C. (1994). Accuracy of planar reaching movements. *Experimental Brain Research*, 99:97–111.

Gorostiza, J., Barber, R., Khamis, A., Malfaz, M., Pacheco, R., Rivas, R., Corrales, A., Delgado, E., and Salichs, M. (2006). Multimodal Human-Robot Interaction Framework for a Personal Robot. In *Proceedings of the 15th IEEE International Symposium on Robot and Human Interactive Communication.*

Gouaillier, D., Hugel, V., Blazevic, P., Kilner, C., Monceaux, J., Lafourcade, P., Marnier, B., Serre, J., and Maisonnier, B. (2009). Mechatronic design of NAO humanoid. In *Proceedings of the IEEE International Conference on Robotics and Automation,* pages 769–774.

Gullberg, M. (1998). *Gesture as a communication strategy in second language discourse: A study of learners of French and Swedish.* Lund University Press, Lund.

Gullberg, M., de Bot, K., and Volterra, V. (2008). Gestures and some key issues in the study of language development. *Gesture,* 8(2):149–179.

Gullberg, M. and Holmqvist, K. (1999). Keeping an Eye on Gestures: Visual Perception of Gestures in Face-to-Face Communication. *Pragmatics and Cognition,* 7:35–63.

Habets, B., Kita, S., Shao, Z., Özyürek, A., and Hagoort, P. (2011). The Role of Synchrony and Ambiguity in Speech: Gesture Integration During Comprehension. *Journal of Cognitive Neuroscience. Advance Online Publication,* 23(8):1845–1854.

Hadar, U. (1989). Two Types of Gesture and their Role in Speech Production. *Journal of Language and Social Psychology,* 8:221–228.

Hadar, U. and Butterworth, B. (1997). Iconic gestures, imagery and word retrieval in speech. *Semiotica,* 115:147–172.

Haggard, P., Hutchinson, K., and Stein, J. (1995). Patterns of coordinated multi-joint movement. *Experimental Brain Research,* 107:254–266.

Ham, J., Bokhorst, R., Cuijpers, R. H., van der Pol, D., and Cabibihan, J.-J. (2011). Making Robots Persuasive: The Influence of Combining Persuasive Strategies (Gazing and Gestures) by a Storytelling Robot on Its Persuasive

Power. In *Proceedings of the International Conference on Social Robotics*, pages 71–83.

Hartmann, B., Mancini, M., and Pelachaud, C. (2002). GRETA: Formational Parameters and Adaptive Prototype Instantiation for MPEG-4 Compliant Gesture Synthesis. In *Proceedings of Computer Animation, IEEE Computer Society*, pages 111–119, Geneva, Switzerland.

Hartmann, B., Mancini, M., and Pelachaud, C. (2005). Implementing Expressive Gesture Synthesis for Embodied Conversational Agents. *Gesture in Human-Computer Interaction and Simulation*.

Haslam, N., Bain, P., Loughnan, S., and Kashima, Y. (2008). Attributing and Denying Humanness to Others. *European Review of Social Psychology*, 19:55–85.

Heloir, A. and Kipp, M. (2010). Real-Time Animation of Interactive Agents: Specification and Realization. *Applied Artificial Intelligence*, 24(6):510–529.

Heylen, D., van Es, I., Nijholt, A., and van Dijk, B. (2002). Experimenting with the Gaze of a Conversational Agent. In *Proceedings International CLASS Workshop on Natural, Intelligent and Effective Interaction in Multimodal Dialogue Systems*, pages 93–100.

Hirose, M., Haikawa, Y., Takenaka, T., and Hirai, K. (2001). Development of humanoid robot ASIMO. In *Workshop Proceedings of the IEEE/RSJ International Conference on Intelligent Robots and Systems*.

Hiyakumoto, L., Prevost, S., and Cassell, J. (1997). Semantic and Discourse Information for Text-to-Speech Intonation. In Alter, K., Pirker, H., and Finkler, W., editors, *Concept to Speech Generation Systems. Proceedings of the ACL Workshop on Concept-to-Speech Generation*, pages 47–56.

Hollerbach, J. M. and Atkeson, C. G. (1987). Deducing planning variables from experimental arm trajectories: pitfalls and possibilities. *Biological Cybernetics*, 56:279–292.

Hostetter, A. B. (2011). When Do Gestures Communicate? A Meta-Analysis. *Psychological Bulletin*, 137(2):297–315.

Hostetter, A. B. and Alibali, M. W. (2008). Visible Embodiment: Gestures as Simulated Action. *Psychonomic Bulletin and Review*, 15(3):495–514.

Itoh, K., Matsumoto, H., Zecca, M., Takanobu, H., Roccella, S., Carrozza, M., Dario, P., and Takanishi, A. (2004). Various Emotional Expressions with Emotion Expression Humanoid Robot WE-4RII. In *Proceedings of the 1st IEEE Technical Exhibition Based Conference on Robotics and Automation Proceedings*, pages 35–36.

Iverson, J. M. and Goldin-Meadow, S. (1997). What's communication got to do with it? Gesture in children blind from birth. *Developmental Psychology*, 33(3):453–467.

Iverson, J. M. and Goldin-Meadow, S. (1998). *The Nature and Functions of Gesture in Children's Communication: New Directions for Child and Adolescent Development*. Jossey Bass, San Francisco.

Jordan, M. I. and Wolpert, D. M. (1999). Computational motor control. In Gazzaniga, M., editor, *The Cognitive Neurosciences*, pages 601–620. MIT Press, Cambridge, MA, USA.

Kawato, M. (1999). Internal models for motor control and trajectory planning. *Current Opinion in Neurobiology*, 9:718–727.

Kawato, M., Maeda, Y., Uno, Y., and Suzuki, R. (1990). Trajectory formation of arm movement by cascade neural network model based on minimum torque-change criterion. *Biological Cybernetics*, 62:275–288.

Keele, S. W. and Posner, M. I. (1968). Processing visual feedback in rapid movements. *Journal of Experimental Psychology*, 77:155–158.

Kelly, S. D., Barr, D. J., Church, R. B., and Lynch, K. (1999). Offering a Hand to Pragmatic Understanding: The Role of Speech and Gesture in Comprehension and Memory. *Journal of Memory and Language*, 40(4):577–592.

Kelly, S. D., Kravitz, C., and Hopkins, M. (2004). Neural correlates of bimodal speech and gesture comprehension. *Brain and Language*, 89(1):253–260.

Kendon, A. (1972). Some relationships between body motion and speech. In Siegman, A. W. and Pope, B., editors, *Studies in Dyadic Communication*, pages 177–216. Pergamon Press, New York.

Kendon, A. (1980). Gesticulation and Speech: Two Aspects of the Process of Utterance. In Key, M. R., editor, *The Relationship of Verbal and Non-verbal Communication*, pages 207–227. Mouton, Den Haag.

Kendon, A. (1986). Current Issues in the Study of Gesture. In Nespoulous, J.-L., Perron, P., and Lecours, A. R., editors, *Biological Foundations of Gestures: Motor and Semiotic Aspects*, pages 23–47. Lawrence Erlbaum Associates, Hillsdale, New Jersey.

Kendon, A. (1988). How Gestures Can Become Like Words. In Poyatos, F., editor, *Crosscultural Perspectives in Nonverbal Communication*, pages 131–141. C. J. Hogrefe, Toronto.

Kendon, A. (1994). Do gestures communicate? A review. *Research on Language and Social Interaction*, 27(3):175–200.

Kendon, A. (1997). Gesture. *Annual Review of Anthropology*, 26:109–128.

Kendon, A. (2004). Gesture: Visible Action as Utterance. *Gesture*, 6(1):119–144.

Kendon, A. (2007). Some Topics in Gesture Studies. In Esposito, A., Bratanic, M., Keller, E., and Marinaro, M., editors, *Fundamentals of Verbal and Nonverbal Communication and the Biometric Issue*, pages 3–19. IOS Press, Amsterdam.

Kim, H., Kwak, S. S., and Kim, M. (2008). Personality Design of Sociable Robots by Control of Gesture Design Factors. In *Proceeding of the 17th IEEE International Symposium on Robot and Human Interactive Communication*, pages 494–499.

Kipp, M. and Gebhard, P. (2008). IGaze: Studying reactive gaze behavior in semi-immersive human-avatar interactions. In *Proceedings of the 8th International Conference on Intelligent Virtual Agents*, volume 5208, pages 191–199. Springer.

Kipp, M., Neff, M., Kipp, K. H., and Albrecht, I. (2007). Towards Natural Gesture Synthesis: Evaluating Gesture Units in a Data-Driven Approach to Gesture

Synthesis. In *Proceedings of the 7th International Conference on Intelligent Virtual Agents*, pages 15–28, Berlin, Heidelberg. Springer.

Kita, S. (1990). *The Temporal Relationship between Gesture and Speech: A Study of Japanese-English Bilinguals.* Unpublished MA Thesis, Department of Psychology, University of Chicago.

Kita, S. (2000). How Representational Gestures Help Speaking. In McNeill, D., editor, *Language and Gesture*, pages 162–185. Cambridge University Press, Cambridge, UK.

Kita, S. (2009). Cross-cultural variation of speech-accompanying gesture: A review. *Language and Cognitive Processes*, 24(2):145–167.

Kita, S. and Özyürek, A. (2003). What does Cross-Linguistic Variation in semantic coordination of speech and gesture reveal? Evidence for an interface representation of spatial thinking and speaking. *Journal of Memory and Language*, 48:16–32.

Kita, S., van Gijn, I., and van der Hulst, H. (1998). Movement phases in signs and co-speech gestures, and their transcription by human coders. In Wachsmuth, I. and Fröhlich, M., editors, *Gesture and Sign Language in Human-Computer Interaction*, volume 1371 of *Lecture Notes in Computer Science*, pages 23–35. Springer, Berlin, Heidelberg.

Kopp, S. (2003). *Synthese und Koordination von Sprache und Gestik für virtuelle multimodale Agenten*, volume 265 of *DISKI*. Infix Akademische Verlagsgesellschaft.

Kopp, S., Bergmann, K., and Wachsmuth, I. (2008). Multimodal Communication from Multimodal Thinking – Towards an Integrated Model of Speech and Gesture Production. *Semantic Computing*, 2(1):115–136.

Kopp, S., Krenn, B., Marsella, S., Marshall, A. N., Pelachaud, C., Pirker, H. H., Thórisson, K. R., and Vilhjálmsson, H. (2006). Towards a Common Framework for Multimodal Generation: The Behavior Markup Language. In J. Gratch, J., Young, M. R., Aylett, R. S., Ballin, D., and Olivier, P., editors, *Proceedings of the 6th International Conference on Intelligent Virtual Agents*, volume 4133 of *Lecture Notes in Computer Science*, pages 205–217. Springer.

Kopp, S., Tepper, P., and Cassell, J. (2004). Towards Integrated Microplanning of Language and Iconic Gesture for Multimodal Output. In *Proceedings of the International Conference on Multimodal Interfaces*, pages 97–104. ACM Press.

Kopp, S. and Wachsmuth, I. (2002). Model-based Animation of Coverbal Gesture. In *Proceedings of Computer Animation 2002*, pages 252–257. IEEE Press.

Kopp, S. and Wachsmuth, I. (2004). Synthesizing Multimodal Utterances for Conversational Agents. *Computer Animation and Virtual Worlds*, 15(1):39–52.

Krämer, N., Simons, N., and Kopp, S. (2007). The Effects of an Embodied Conversational Agents Nonverbal Behavior on Users Evaluation and Behavioral Mimicry. In *Proceedings of the 7th Conference on Intelligent Virtual Agents*, volume 4722, pages 238–251. Springer.

Krämer, N., Tietz, B., and Bente, G. (2003). Effects of Embodied Interface Agents and Their Gestural Activity. In Rist, T., Aylett, R., Ballin, D., and Rickel, J., editors, *Intelligent Virtual Agents*, volume 2792 of *Lecture Notes in Computer Science*, pages 292–300. Springer, Berlin, Heidelberg.

Kranstedt, A., Kopp, S., and Wachsmuth, I. (2002). MURML: A Multimodal Utterance Representation Markup Language for Conversational Agents. In *Proceedings of the AAMAS 2002 Workshop on Embodied Conversational Agents - Let's Specify and Evaluate Them*, Bologna, Italy.

Krauss, R., Chen, Y., and Gottesman, R. (2000). Lexical gestures and lexical access: A process model. In McNeill, D., editor, *Language and Gesture*, pages 261–283. Cambridge University Press, Cambridge, UK.

Krauss, R. M., Chen, Y., and Chawla, P. (1996). Nonverbal behavior and nonverbal communication: What do conversational hand gestures tell us? In Zanna, M., editor, *Advances in Experimental Social Psychology*, pages 389–450. Academic Press, Tampa.

Krauss, R. M., Dushay, R. A., Chen, Y., and Rauscher, F. (1995). The Communicative Value of Conversational Hand Gestures. *Journal of Experimental Social Psychology*, 31(6):533–552.

Krauss, R. M., Morrel-Samuels, P., and Colasante, C. (1991). Do Conversational Hand Gestures Communicate? *Journal of Personality and Social Psychology*, 61(5):743–754.

Latash, M. (1993). *Control of Human Movement*. Human Kinetics Publishers, Champaign, IL, USA.

Latash, M. (2008). *Neurophysiological Basis of Movement. Second Edition*. Human Kinetics, Champaign, IL, USA.

Latash, M. L. (1996). The Bernstein Problem: How does the CNS make its choices? In Latash, M. L. and Turvey, M. T., editors, *Dexterity and its Development*, pages 277–303. Lawrence Erlbaum.

Latash, M. L., Scholz, J. P., and Schöner, G. (2007). Toward a New Theory of Motor Synergies. *Motor Control*, 11(3):276–308.

Le, Q. A., Hanoune, S., and Pelachaud, C. (2011). Design and implementation of an expressive gesture model for a humanoid robot. In *Proceedings of the 11th IEEE-RAS International Conference on Humanoid Robots*, pages 134–140, Bled, Slovenia.

Lee, J. and Marsella, S. (2006). Nonverbal Behavior Generator for Embodied Conversational Agents. In *Proceedings of the 6th International Conference on Intelligent Virtual Agents*, pages 243–255, Marina del Rey, CA. Springer.

Lee, J., Wang, Z., and Marsella, S. (2010). Evaluating Models of Speaker Head Nods for Virtual Agents. In *Proceedings of the 9th International Conference on Autonomous Agents and Multiagent Systems*, pages 1257–1264.

Levelt, W. (1989). *Speaking*. MIT Press, Cambridge, MA.

Levelt, W. J. M., Richardson, G., and La Heij, W. (1985). Pointing and Voicing in Deictic Expressions. *Journal of Memory and Language*, 24(2):133–164.

Li, J., Chignell, M., Mizobuchi, S., and Yasumura, M. (2009). Emotions and Messages in Simple Robot Gestures. In *Proceedings of the 13th International Conference on Human-Computer Interaction. Part II: Novel Interaction Methods and Techniques*, pages 331–340, Berlin, Heidelberg. Springer.

Liégeois, A. (1977). Automatic Supervisory Control of the Configuration and Behavior of Multibody Mechanisms. *IEEE Transactions on Systems, Man and Cybernetics*, 7(12):868–871.

Loehr, D. (2007). Aspects of rhythm in gesture and speech. *Gesture*, 7(2):179–214.

Macdorman, K. and Ishiguro, H. (2006). The Uncanny Advantage of Using Androids in Cognitive and Social Science Research. *Interaction Studies*, 7(3):297–337.

MacKenzie, I. S. (1992). Fitts' law as a research and design tool in human-computer interaction. *Human-Computer Interaction*, 7(1):91–139.

Maslovat, D., Hayes, S., Horn, R., and Hodges, N. (2010). Motor Learning Through Observation. In Elliott, D. and Khan, M., editors, *Vision and Goal-Directed Movement: Neurobehavioural Perspectives*, pages 315–340. Human Kinetics, Champaign, IL, USA.

Mataric, M. J. and Pomplun, M. (1998). Fixation Behavior in Observation and Imitation of Human Movement. *Cognitive Brain Research*, 7(2):191 – 202.

McClave, E. (1994). Gestural Beats: The Rhythm Hypothesis. *Journal of Psycholinguistic Research*, 23:45–66.

McNeill, D. (1992). *Hand and Mind: What Gestures Reveal about Thought*. University of Chicago Press, Chicago.

McNeill, D. (2000). Introduction. In *Language and Gesture*, pages 1–10. Cambridge University Press, Cambridge, UK.

McNeill, D. (2005). *Gesture and Thought*. University of Chicago Press, Chicago.

McNeill, D., Cassell, J., and McCullough, K.-E. (1994). Communicative Effects of Speech-Mismatched Gestures. *Research on Language & Social Interaction*, 27(3):223–237.

McNeill, D. and Duncan, S. D. (2000). Growth points in thinking-for-speaking. In *Language and Gesture*, pages 141–161. Cambridge University Press, Cambridge, UK.

McNeill, D. and Levy, E. (1982). Conceptual Representations in Language Activity and Gesture. In Jarvella, R. and Klein, W., editors, *Speech, Place and Action*, pages 271–295. John Wiley & Sons, Chichester.

Mead, R., Wade, E., Johnson, P., St. Clair, A. B., Chen, S., and Matarić, M. J. (2010). An Architecture for Rehabilitation Task Practice in Socially Assistive Human-Robot Interaction. In *Proceedings of the 19th IEEE International Symposium in Robot and Human Interactive Communication*, pages 404–409, Viareggio, Italy.

Melinger, A. and Kita, S. (2004). When Input and Output Diverge: Mismatches in Gesture, Speech and Image. In *Proceedings of the 26th Annual Meeting of the Cognitive Science Society*, Chicago.

Melinger, A. and Levelt, W. (2004). Gesture and the Communicative Intention of the Speaker. *Gesture*, 4:119–141.

Minato, T., Shimada, M., Ishiguro, H., and Itakura, S. (2004). Development of an Android Robot for Studying Human-Robot Interaction. *Innovations in Applied Artificial Intelligence*, pages 424–434.

Miyashita, T., Shinozawa, K., and Hagita, N. (2006). Gesture Translation for Heterogeneous Robots. In *Proceedings of 6th IEEE-RAS International Conference on Humanoid Robots*, pages 462–467.

Morasso, P. (1981). Spatial control of arm movements. *Experimental Brain Research*, 42:223–227.

Morasso, P. (1986). Trajectory Formation. In Morasso, P. and Tagliasco, V., editors, *Human Movement Understanding*, pages 7–58. Elsevier Science Publishers B.V., Amsterdam, The Netherlands.

Mori, M. (1970). The uncanny valley (K. F. MacDorman & T. Minato, Trans.). *Energy*, 7(4):33–35.

Murdock, B. B. (1962). The Serial Position Position Effect of Free Recall. *Journal of Experimental Psychology*, 64(5):482–488.

Mutlu, B., Shiwa, T., Kanda, T., Ishiguro, H., and Hagita, N. (2009). Footing in human-robot conversations: how robots might shape participant roles using gaze cues. In *Proceedings of the 4th ACM/IEEE International Conference on Human-Robot Interaction*, pages 61–68.

Narahara, H. and Maeno, T. (2007). Factors of Gestures of Robots for Smooth Communication with Humans. In *Proceedings of the 1st international Conference on Robot Communication and Coordination*, pages 44:1–44:4, Piscataway, NJ, USA. IEEE Press.

Neff, M., Wang, Y., Abbott, R., and Walker, M. (2010). Evaluating the Effect of Gesture and Language on Personality Perception in Conversational Agents. In *Proceedings of the 10th International Conference on Intelligent Virtual Agents*, pages 222–235, Berlin, Heidelberg. Springer.

Nehaniv, C. L. and Dautenhahn, K. (2001). Like me? - Measures of Correspondence and Imitation. *Cybernetics & Systems: An International Journal*, 32(1-2):11–51.

Nehaniv, C. L. and Dautenhahn, K. (2002). The Correspondence Problem. In Dautenhahn, K. and Nehaniv, C. L., editors, *Imitation in Animals and Artifacts*, pages 41–61. MIT Press, Cambridge, MA, USA.

Ng-Thow-Hing, V., Luo, P., and Okita, S. (2010). Synchronized Gesture and Speech Production for Humanoid Robots. In *Proceedings of the IEEE/RSJ International Conference on Intelligent Robots and Systems*, pages 4617–4624.

Noma, T., Zhao, L., and Badler, N. I. (2000). Design of a Virtual Human Presenter. *IEEE Computer Graphics and Applications*, 20(4):79–85.

O'Brien, K., Sutherland, J., Rich, C., and Sidner, C. L. (2011). Collaboration with an Autonomous Humanoid Robot: A Little Gesture Goes a Long Way. In *Proceedings of the 6th ACM/IEEE International Conference on Human-Robot Interaction*, pages 215–216.

Okuno, Y., Kanda, T., Imai, M., Ishiguro, H., and Hagita, N. (2009). Providing Route Directions: Design of Robot's Utterance, Gesture, and Timing. In *Proceedings of the 4th ACM/IEEE International Conference on Human-Robot Interaction*, pages 53–60. ACM.

Özyürek, A., Willems, R. M., Kita, S., and Hagoort, P. (2007). On-line integration of semantic information from speech and gesture: Insights from event-related brain potentials. *Journal of Cognitive Neuroscience*, 19(4):605–616.

Pammi, S., Charfuelan, M., and Schröder, M. (2010). Multilingual Voice Creation Toolkit for the MARY TTS Platform. In *LREC*, pages 3750 – 3756.

Park, E., Kim, K. J., and del Pobil, A. P. (2011). The Effects of Robot's Body Gesture and Gender in Human-Robot Interaction. In *Internet and Multimedia Systems and Applications: Human-Computer Interaction*, volume 747. ACTA Press, Washington, DC, USA.

Peirce, C. (1960). *Collected Papers of Charles Sanders Peirce*, volume 2. Harvard University Press, Cambridge, MA.

Pollard, N., Hodgins, J., Riley, M., and Atkeson, C. (2002). Adapting Human Motion for the Control of a Humanoid Robot. In *Proceedings of International Conference on Robotics and Automation*, pages 1390–1397.

Prillwitz, S., Leven, R., Zienert, H., Hanke, T., and Henning, J. (1989). *HamNoSys Version 2.0. Hamburg Notation System for Sign Languages: An Introductory Guide*. Signum, Hamburg.

Rauscher, F. H., Krauss, R. M., and Chen, Y. (1996). Gesture, Speech, and Lexical Access: The Role of Lexical Movements in the Processing of Speech. *Psychological Science*, 7(4):226–231.

Reiter, E. and Dale, R. (2000). *Building Natural Language Generation Systems*. Cambridge Univ. Press.

Riek, L. D., Rabinowitch, T., Bremner, P., Pipe, A. G., Fraser, M., and Robinson, P. (2010). Cooperative Gestures: Effective Signaling for Humanoid Robots. In *Proceeding of the 5th ACM/IEEE International Conference on Human-Robot Interaction*, pages 61–68.

Rimé, B. (1982). The elimination of visible behaviour from social interactions: Effects on verbal, nonverbal and interpersonal variables. *European Journal of Social Psychology*, 12(2):113–129.

Rimé, B. and Schiaratura, L. (1991). Gesture and Speech. In Feldman, R. and Rimé, R., editors, *Fundamentals of Nonverbal Behavior*, pages 239–281. Press Syndicate of the University of Cambridge, New York.

Rosenbaum, D. A. (2002). Motor Control. In Pashler, H. and Yantis, S., editors, *Stevens Handbook of Experimental Psychology. Third Edition*, pages 315–339. John Wiley & Sons, Inc.

Roth, W. (2002). From action to discourse: The bridging function of gestures. *Cognitive Systems Research*, 3:535–554.

Salem, M. (2011). Coneptual Motorics – Generation and Analysis of Communicative Robot Gesture. In *Proceedings of the 2011 Human-Robot Interaction Pioneers Workshop*, pages 78–79.

Salem, M., Eyssel, F., Rohlfing, K., Kopp, S., and Joublin, F. (2011a). Effects of Gesture on the Perception of Psychological Anthropomorphism: A Case Study with a Humanoid Robot. In *Proceedings of the Third International Conference on Social Robotics*, volume 7072 of *Lecture Notes in Artificial Intelligence*, pages 31–41. Springer.

Salem, M., Kopp, S., Wachsmuth, I., and Joublin, F. (2009). Towards Meaningful Robot Gesture. In Ritter, H., Sagerer, G., Dillmann, R., and Buss, M., editors, *Human Centered Robot Systems*, volume 6 of *Cognitive Systems Monographs*, pages 173–182. Springer.

Salem, M., Kopp, S., Wachsmuth, I., and Joublin, F. (2010a). Generating Multi-Modal Robot Behavior Based on a Virtual Agent Framework. In *Proceedings of the IEEE International Conference on Robotics and Automation: Workshop on Interactive Communication for Autonomous Intelligent Robots*, pages 23–25.

Salem, M., Kopp, S., Wachsmuth, I., and Joublin, F. (2010b). Generating Robot Gesture Using a Virtual Agent Framework. In *Proceedings of the IEEE/RSJ International Conference on Intelligent Robots and Systems*, pages 3592–3597.

Salem, M., Kopp, S., Wachsmuth, I., and Joublin, F. (2010c). Towards an Integrated Model of Speech and Gesture Production for Multi-Modal Robot Behavior. In *Proceedings of the 19th IEEE International Symposium on Robot and Human Interactive Communication*, pages 649–654.

Salem, M., Kopp, S., Wachsmuth, I., and Joublin, F. (2011b). A Multimodal Scheduler for Synchronized Humanoid Robot Gesture and Speech. In *Book of Extended Abstracts of the 9th International Gesture Workshop*, pages 64–67.

Salem, M., Kopp, S., Wachsmuth, I., Rohlfing, K., and Joublin, F. (2012). Generation and Evaluation of Communicative Robot Gesture. *International Journal of Social Robotics, Special Issue on Expectations, Intentions, and Actions*, 4(2):201–217.

Salem, M., Rohlfing, K., Kopp, S., and Joublin, F. (2011c). A Friendly Gesture: Investigating the Effect of Multimodal Robot Behavior in Human-Robot Interaction. In *Proceedings of the 20th IEEE International Symposium on Robot and Human Interactive Communication*, pages 247–252.

Salichs, M., Barber, R., Khamis, A., Malfaz, M., Gorostiza, J., Pacheco, R., Rivas, R., Corrales, A., Delgado, E., and Garcia, D. (2006). Maggie: A Robotic Platform for Human-Robot Social Interaction. In *IEEE International Conference on Robotics, Automation and Mechatronics*, pages 1–7.

Saltzman, E. (1979). Levels of sensorimotor representation. *Journal of Mathematical Psychology*, 20(2):91 – 163.

Saygin, A., Chaminade, T., Ishiguro, H., Driver, J., and Frith, C. (2011). The Thing That Should Not Be: Predictive Coding and the Uncanny Valley in Perceiving Human and Humanoid Robot Actions. *Social Cognitive and Affective Neuroscience*.

Schaal, S., Ijspeert, A., and Billard, A. (2003). Computational Approaches to Motor Learning by Imitation. *Philosophical Transaction of the Royal Society of London: Series B, Biological Sciences*, 358(1431):537–547.

Schegloff, E. A. (l984). On Some Gestures' Relation to Talk. In Atkinson, J. M. and Heritage, J., editors, *Structures of Social Action*, pages 266–298. Cambridge University Press, Cambridge, UK.

Schmidt, R. A. (1982). The motor program. In Kelso, J. A. S., editor, *Human Motor Behavior: An Introduction*, pages 187–218. Erlbaum, Hillsdale, NJ, USA.

Schröder, M. (2004). *Speech and Emotion Research: An Overview of Research Frameworks and a Dimensional Approach to Emotional Speech Synthesis*, volume 7 of *PHONUS*. Infix Akademische Verlagsgesellschaft, Saarland University, Germany.

Schröder, M. and Trouvain, J. (2003). The German Text-to-Speech Synthesis System MARY: A Tool for Research, Development and Teaching. *International Journal of Speech Technology*, pages 365–377.

Shams, L. and Seitz, A. (2008). Benefits of multisensory learning. *Trends in Cognitive Sciences*, 12(11):411–417.

Sidner, C., Lee, C., and Lesh, N. (2003). The Role of Dialog in Human Robot Interaction. In *International Workshop on Language Understanding and Agents for Real World Interaction*.

Sowa, T., Kopp, S., Duncan, S., McNeill, D., and Wachsmuth, I. (2008). Implementing a non-modular theory of language production in an embodied conversational agent. In Wachsmuth, I., Lenzen, M., and Knoblich, G., editors, *Embodied Communication*, pages 425–449. Oxford University Press, Oxford.

Stein, T. (2010). *Computational Motor Control of Human Movements. Dissertation*. Technische Universität Darmstadt [online]. http://tuprints.ulb.tu-darmstadt.de/2095.

Sugiura, H., Gienger, M., Janssen, H., and Goerick, C. (2006). Real-Time Self Collision Avoidance for Humanoids by means of Nullspace Criteria and Task Intervals. In *Proccedings of the 6th IEEE-RAS International Conference on Humanoid Robots*, pages 575–580.

Sugiura, H., Gienger, M., Janssen, H., and Goerick, C. (2007). Real-Time Collision Avoidance with Whole Body Motion Control for Humanoid Robots. In *Proceedings of the IEEE/RSJ International Conference on Intelligent Robots and Systems*, pages 2053–2058.

Sugiura, H., Janßen, H., and Goerick, C. (2009). Instant Prediction for Reactive Motions with Planning. In *Proceedings of the 2009 IEEE/RSJ International Conference on Intelligent Robots and Systems*, pages 5475–5480, Piscataway, NJ, USA. IEEE Press.

Sugiyama, O., Kanda, T., Imai, M., Ishiguro, H., and Hagita, N. (2007). Natural Deictic Communication with Humanoid Robots. In *Proceedings of the IEEE International Conference on Intelligent Robots and Systems*, pages 1441–1448.

Thiebaux, M., Marsella, S., Marshall, A. N., and Kallmann, M. (2008). SmartBody: Behavior Realization for Embodied Conversational Agents. In *Proceedings of the 7th International Joint Conference on Autonomous Agents and Multiagent Systems*, volume 1, pages 151–158, Richland, SC.

Tokuda, K., Zen, H., and Black, A. (2004). HMM-based Approach to Multilingual Speech Synthesis. In Narayanan, S. and Alwan, A., editors, *Text to Speech Synthesis: New Paradigms and Advances*. Prentice Hall.

Tolani, D., Goswami, A., and Badler, N. I. (2000). Real-time inverse kinematics techniques for anthropomorphic limbs. *Graph. Models Image Process.*, 62:353–388.

Turvey, M. T. (1990). Coordination. *American Psychologist*, 45:938–953.

Väänänen, K. and Böhm, K. (1993). Gesture-driven Interaction as a Human Factor in Virtual Environments - An Approach with Neural Networks. In Jones, H., Gigante, M. A., and Earnshaw, R. A., editors, *Virtual Reality Systems*, pages 93–106. Academic Press Ltd., London.

van Welbergen, H., Reidsma, D., Ruttkay, Z., and Zwiers, J. (2010). Elckerlyc - A BML Realizer for continuous, multimodal interaction with a Virtual Human. *Journal on Multimodal User Interfaces*, 3(4):271–284.

Vilhjálmsson, H., Cantelmo, N., Cassell, J., Chafai, N. E., Kipp, M., Kopp, S., Mancini, M., Marsella, S., Marshall, A. N., Pelachaud, C., Ruttkay, Z., Thorisson, K. R., van Welbergen, H., and van der Werf, R. J. (2007). The Behavior Markup Language: Recent Developments and Challenges. In Pelachaud, C., Martin, J.-C., Andre, E., Collet, G., Karpouzis, K., and Pele, D., editors, *Proceedings of the 7th International Conference on Intelligent Virtual Agents*, volume 4722 of *Lecture Notes in Computer Science*, pages 99–111, Berlin. Springer.

Von Holst, E. and Mittelstaedt, H. (1950). Das Reafferenzprinzip. Wechselwirkungen zwischen Zentralnervensystem und Peripherie. *Die Naturwissenschaften*, 37(20):464–476.

Wachsmuth, I. and Kopp, S. (2002). Lifelike Gesture Synthesis and Timing for Conversational Agents. In Wachsmuth, I. and Sowa, T., editors, *Gesture and Sign Language in Human-Computer Interaction*, LNAI 2298, pages 120–133, Berlin. Springer.

Wachsmuth, I. and Salem, M. Body movements in robotics. In Müller, C., Cienki, A., Fricke, E., and McNeill, D., editors, *Handbook "Body-Language-Communication"*. De Gruyter Mouton, Berlin, New York. To appear.

Wohlschläger, A., Gattis, M., and Bekkering, H. (2003). Action generation and action perception in imitation: an instance of the ideomotor principle. *Philosophical Transaction of the Royal Society of London: Series B, Biological Sciences*, 358(1431):501–515.

Wolpert, D. and Flanagan, J. (2001). Motor prediction. *Current Biology*, 11(18):R729–R732.

Wolpert, D. M. (1997). Computational approaches to motor control. *Trends in Cognitive Sciences*, 1(7):209–216.

Wolpert, D. M., Ghahramani, Z., and Jordan, M. I. (1995). Are arm trajectories planned in kinematic or dynamic coordinates? An adaptation study. *Experimental Brain Research*, 103(3):460–470.

Wundt, W. M. (1900/1973). *The Language of Gestures*. Mouton, Den Haag.

Zeltzer, D. (1982). Motor Control Techniques for Figure Animation. *IEEE Computer Graphics Applications*, 2(9):53–59.